ABOUT THE AUTHOR

BRUCE STERLING is the author of nine novels, three of which were selected as *New York Times* Notable Books of the Year. *The Difference Engine,* co-written with William Gibson, was a national bestseller. He has also published three short-story collections and one nonfiction book, *The Hacker Crackdown.* He edited the anthology *Mirrorshades* and has written for many magazines, including *Newsweek, Fortune, Harper's, Details, Whole Earth Review,* and *Wired,* where he has been a contributing writer since its conception. In 1999, he won the Hugo Award in the short-story category. He lives in Austin, Texas.

"[Sterling] writes in a plain, conversational style, which he interrupts with striking, apothegmatic formulations, i.e., sound-bites. . . . Some of these formulations sound almost too flip, but nearly all of them also seem correct or close to it. . . . Entertaining and stimulating."   —*The Washington Post*

"[*Tomorrow Now*] maintains Sterling's customary quick pace. His other strong points are also in evidence: a truly global world view; clear-eyed assessment of new technologies and their impacts; and a wit as tart as a food processor full of lemon rinds. . . . A canny guesser, Sterling, with his entertaining style and unorthodox mindset, is a great guide to tomorrow. You may or may not agree with his conclusions, but you'll be glad you read this book."   —*The Seattle Times*

"Sterling is in the business of holding up a mirror to today and locating tomorrow somewhere in the reflection. When he gets the angle just right, the effect can be dazzling. . . . It's a semi-scholarly work of nonfiction by a man who takes the future seriously. . . . Funny, fascinating . . . "Stage Seven: Mere Oblivion" is explosive apocalyptic stuff."   —*The San Diego Union-Tribune*

"The most striking aspect of this volume is its optimism. . . . Sterling doesn't settle any issues, but he makes you think harder about the literal meaning of life and death. Throughout, Sterling infuses deep topics and moral quandaries with wit and insight."   —*San Jose Mercury News*

"Sterling provides an informative, straightforward and relatively jargon-free discussion of critical issues. . . . He has interesting things to say about education, politics, the economy and terrorism . . . His examples are drawn from his wide reading, as well as his personal experiences as a writer and father. And he doesn't hide his own quirky interests and obsessions."

—*The Oregonian*

"Transforming hype into epigrams and forebodings into witty deconstructions of today's moral panics . . . A tour de force . . . Well above ephemeral trend-spotting, Sterling's high-IQ futurism is sure to be devoured by hackers and the remaining Silicon Valley CEOs."

—*Kirkus Reviews*

"Sterling's breezy tone and insightful speculations reposition this 'cyberpunk' hero as a fun hybrid of Robert Kaplan and Faith Popcorn, ready to join the punditocracy and reach out to a broader readership."

—*Publishers Weekly*

TOMORROW NOW

TOMORROW NOW

ENVISIONING THE NEXT FIFTY YEARS

BRUCE STERLING

RANDOM HOUSE TRADE PAPERBACKS NEW YORK

2003 Random House Trade Paperback Edition

Copyright © 2002, 2003 by Bruce Sterling

All rights reserved under International and Pan-American Copyright
Conventions. Published in the United States by Random House Trade
Paperbacks, an imprint of The Random House Publishing Group,
a division of Random House, Inc., New York, and simultaneously
in Canada by Random House of Canada Limited, Toronto.

RANDOM HOUSE TRADE PAPERBACKS and colophon are trademarks
of Random House, Inc.

This work was originally published in slightly different form in hardcover
by Random House in 2002.

Library of Congress Cataloging-in-Publication Data
Sterling, Bruce.
    Tomorrow now: envisioning the next fifty years / Bruce Sterling.
       p.   cm.
    ISBN 0-8129-6976-6
    1. Social prediction. 2. Twenty-first century—Forecasts. I. Title.
  HM901 .S74 2002
    303.49—dc21   2002024801

Random House website address: www.atrandom.com

Printed in the United States of America

9 8 7 6 5 4 3 2

Book design by Casey Hampton

CONTENTS

context of the passage of time. Futurism teaches how to recognize profoundly changing circumstances. It offers paths forward that might have reasonable prospects of success. It offers humanity some fresh mistakes.

We humans can never tackle time in its raw form. We can tackle only our perceptions of time: the language, the mindset, the sensibility, and the period frame of mind. Successful futurism assembles evidence of trends to aim at paradigms. Futurism is an art of re-perception. It means recognizing that life will change, must change, and has changed, and it suggests how and why. It shows that old perceptions have lost their validity, while new ones are possible.

Futurism tends to work best in threes—not because time works in threes but because the human mind does. Good futuristic forecasts are like detective work, based in three factors: means, motive, and opportunity.

The means are the driving forces of change. Futurists cluster most eagerly around scientific and technological changes, because their impact is so easy to spot. That's been the case in Western society since the twelfth century, when people first understood the increased human command over forces of nature as progress. The future is no longer a synonym for "progress." We now understand that world-changing trends strike from many angles: technical, social, economic, political, and environmental.

Motives are found in the desires of humankind. People formally announce their intentions for change through po-

In normal circumstances, I'm not the sober, serious futurist that you will see in this work. This is me as a full-blown pundit, a brow-wrinkled journalist who attends the Davos Forum, networks with Californian corporate forecasters, and mourns the tragic loss of the Congressional Office of Technology Assessment. Most of the time I really don't care to work that hard. Because I'm a science fiction writer.

By trade, science fiction writers are pop entertainers. Science fiction can manage just fine without the future. Quite often the future is a drag on our market. The top contemporary earners in science fiction aren't futuristic works, they're fantasies—*Star Wars, Harry Potter,* and *Lord of the Rings*. These mass entertainments have mythic plot structures, with cool special effects and fabulous set designs. People eat that stuff up, and the spin-off toy-collectible market alone is huge—yet it has nothing to do with tomorrow's pressing realities. This book must definitely does; it assaults the future head-on. It is nonfiction, although people who "predict the future" must always tell fictions—even if they happen to tell the truth.

Let me offer a personal example. Back in 1989 I wrote a science fiction story called "We See Things Differently." This twenty-first-century story is narrated by a suicidal jihad Arab terrorist from Egypt who travels to Florida to kill a prominent American political figure by secretly poisoning him with a biowarfare powder.

Come late 2001, and events of that ilk were in headlines all over. In the year 2001 I got quite a lot of e-mail about that story, much of it from people who were genuinely alarmed by my prescience. But I myself wasn't much alarmed. I didn't bother to tell them about the many other science fiction stories I've written that *don't* involve any suicidal Islamic terrorists. For as this book makes clear, although I'm rather thoroughly aware of Islamic terrorists, I don't consider them that big a deal in the long run.

Sci-fi is supposed to be entertainment, yet futurism is a serious enterprise. Let me pull you behind the ol' Wizard of Oz curtain and explain how futurists go about their work.

First, you have to find somebody who'll pay you to do it. This stark reality immediately splits "futurism" into interest groups.

In corporate futurism (which pays the best), you're concerned with new markets and new products. In government futurism (the most dignified), it's about investment in basic R&D, the changing demographics of the political base, and new demands for bureaucratic public service. Military futurism is about new weapons platforms and new security threats: "thinking the unthinkable," as Herman Kahn aptly used to put it. Police futurism (and yes, there are a few police futurists) is about figuring out how to apprehend, prosecute, and jail people involved in wicked activities not yet formally defined as crimes. Ethical futurism is about the moral conundrums posed by possible future actions such as human cloning and genetic alteration. And so forth.

The future is the largest of all possible subjects. It encompasses everything on the far side of the ever-ticking clock. But professional futurists tend to work within narrow parameters. In *Tomorrow Now* I'm combining approaches. Thanks to my independent economic base as a pop entertainer, I can mess with the future in seven different ways all at once.

Someone who could give you a detailed, fully accurate portrait of tomorrow would not be human but a wizard. Such a prophet would also be extremely dangerous and very uncanny, and would have to be arrested. I don't pretend to be one here. At its best, futurism does what history can do. It suggests better decisions about our own actions in the litical party platforms, but those declarations are stilted and phony. Markets supply broader hints about human desire, for people will buy what they want when they can. The most fertile ground for analyzing motives is pop culture—not because pop culture is deep but because it's so shallow. It's where those wishes and longings are most nakedly evident.

Opportunity is the hole in the barriers to change. Opportunity knocks with new materials, new technologies, and new perceptions. The cold war ended when the technological base of one system defeated the other, when people realized that it had an exit. The victor's advantages flowered slowly, but the transition was sudden.

Contradictions and oxymorons often signal future opportunities. They suggest that something formerly unthinkable is within humankind's grasp. A paradox is a gap in competing systems of definition, a potential hole in the status quo, a frontier where the light of comprehension is dawning. "Science fiction" is an oxymoron. That term signals the spark gap between science and literature, an arena where cultural attitudes toward technology can be thrashed out.

The portmanteau word "cyberpunk" pulled the same trick: it jammed together computer science and bohemian rebellion, two areas that were alien to each other yet complementary. Geeks require some trendy glamour for mainstream acceptance, while the trendy live and die through their alternative means of publicity—in this case, computer networks. Thus cyberpunk, an area of pop culture in which I worked happily for years on end. The material there was just great—one rarely saw such utter, stimulating confusion.

Thus the futuristic trio of means, motive, and opportunity. Get those three halfway right, and you can analyze tomorrow's perfect crime. But how to describe the result? What does it mean, and what do we make of it?

Once again the future comes in threes: the upside, the downside, and muddling through. We can even name them after the famous prophets who champion these attitudes. They are Pangloss, Cassandra, and the Insurance Salesman.

Dr. Pangloss is a character from Voltaire's *Candide*. A very learned philosopher, he argues against the existence of evil. For Dr. Pangloss, all damage is collateral damage in a universe that is basically benign. Properly interpreted, every cloud is 90 percent silver lining.

As a guide to the future, Dr. Pangloss has plenty to offer us. People's fears and anxieties about their future are generally worse than undergoing the reality. People also have a marvelous facility for ignoring other people's pain. This is where Dr. Pangloss gets his psychological acumen; as long as it's not happening to *you*, how "evil" can it be? People demonstrate an impressive talent for forgetting even their own suffering. Time heals most wounds. Otherwise, no woman would have a second child; yet every generation breeds new Panglosses, who chuckle indulgently about every grief that befell their parents. Dr. Pangloss may seem a little fatuous, but there's no denying his staying power. He enjoys his own cheery delusions and never has trouble getting out of bed in the morning.

The drawback to a Pangloss approach is that it risks bewilderment, despair, or cynicism after any genuine setback. It's not enough to put a brave face on events; sometimes people do have to be brave.

Cassandra is Homer's famous prophetess of doom, never properly heeded by ignorant scoffers. For Cassandra, the paths of future glory lead but to the grave. Nothing ventured, nothing lost. Things are far worse than they seem. The human enterprise is a ship of fools. Cassandra's greatest benefit to futurology is in facing her harsh truth: that treasured things and circumstances will die and rot. People pay futurists to predict glittering novelties, not to mourn the solemn doom of the obsolete and the overthrown. But holes, defeats, and absences are every bit as much a part of the future as its triumphs and novelties. The twentieth century was the nineteenth with cars—and also without horses.

Cassandra's drawback is that she is melodramatic and eventually boring. She turned down a golden chance to embrace the god Apollo, so her children died young and she was killed by a jealous wife. It's doom for breakfast, lunch, and dinner with Cassandra, for she has no appetite for risk. She deafens the ears that ignore her.

The Insurance Salesman is not from myth or literature. He's from "real life." He cannily predicts that the ol' routine will bumble along much as it always has. Though the flags, names, and labels often change, no "revolution" is ever as big a deal as most revolutionaries let on. No tree ever grows to the sky. The pendulum swings, but the cord rarely snaps. The Insurance Salesman doesn't much care whether you live or you die, but he's willing to place a bet on the odds. Thanks to his close study of decades of actuarial tables, he knows that most people like you do tend to live for a certain amount of time (absent smoking and given seat belts).

The Insurance Salesman is the voice of common sense. His pragmatism deflates Pangloss's hype and bucks up Cassandra's depression. He eats bulls and shoots bears for a living. He considers himself "realistic."

His drawback is that he misses events that are genuinely weird. He can quote the going rate for buggy whips but never for tires. In a catastrophe, insurance companies go broke. Tomorrow's weather is most likely yesterday's weather—but not in all circumstances. Insurance companies are going broke today, in a rote of record-breaking floods, fires, and hurricanes.

The victory condition for futurism is hard to define. One might imagine that it means precisely predicting some important event that nobody ever thought would happen. Then you'd become the Famous Guy Who Predicted That. But dramatic, standout events are rarely truly important. They're symbolic, not determinative.

As the Chinese general Sun Tzu said more than two

millennia ago, the apex of military skill is not a hundred victories in a hundred battles. It's subduing the enemy in such a thorough, silent way that it involves no war. The enemy of Sun Tzu probably doesn't even recognize that Sun Tzu is the enemy. Very likely he considers Sun Tzu a harmless, bumbling, likable figure who needs to be indulged.

In a similar way, the victorious futurist is not a prophet. He or she does not defeat the future but predicts *the present.* Futurism doesn't mean predicting an awesome wonder; rather it means recognizing and describing a small apparent oddity that is destined to become a great commonplace. Cyberpunk involved a lyrical statement of the unthinkable (in the mid-1980s): that someday there would be a world rather like the late 1990s. When the late nineties indeed arrived, cyberpunk was not "predictive"—it was *clichéd.* It was banal and marked as archaic because its authors were feverishly proclaiming certain objects and attitudes that had become background noise. Cyberpunks valorized things that earned a shrug for their corniness fifteen years later. The sense of wonder has a short shelf life.

Nothing obsolesces like "the future." Nothing burns out quite so quickly as a high-tech avant-garde. Technology doesn't glide into the streamlined world of tomorrow. It jolts and limps, all crutches and stilts, just like its ancient patron, the god Hephaestos.

Tomorrow Now is a book about theory and design in an

age of machines. It's an ambitious, sprawling effort in thundering futurist punditry, in the pulsing vein of the futurists I've read and admired over the years: H. G. Wells, Arthur C. Clarke, and Alvin Toffler; Lewis Mumford, Reyner Banham, Peter Drucker, and Michael Dertouzos. Colorful folks who think big, who tackle change on its largest scales, and who are quite unafraid to be loudly wrong.

This book asks the future two questions: *What does it mean?* and *How does it feel?* That is what I would most like to know about the next fifty years. My quest is not for the future's hardware or statistics; it's for its sensibility, its basic attitudes and deep convictions. Its meaning, its feeling.

When I was younger, I had the world ahead of me, a blank page. It's different here, in my involuntarily dignified middle age. I find myself living in the thick of tomorrow. Many things that I merely imagined in the glowing dawn of my career have oozed, more or less, into today's realities. It's not too awfully hard to recognize the cozy, globalized consumers of *Islands in the Net* (1988) in today's malls or in *Fortune* magazine. I've even lived to see my books *The Hacker Crackdown* (1992) and *The Difference Engine* (1990) taught in universities. Taught not as futurism, mind you, but as historical studies.

For a science fiction writer, the passage of time is the gift of Pygmalion. The ancient Greek sculptor, all pitifully enamored of his own genius, saw his most imaginative work brought to life by the goddess Venus. What an irony: that sleek marble masterpiece reduced to being Mrs. Pygmalion! No longer a stunning creative feat of his imagination, icy Galatea became an everyday human being, reduced to living and breathing, laughing and sobbing, dying and rotting. Just like the rest of us.

If I have a core conviction about the future, it's that it isn't made of marble. The future's not real, yet it's no abstraction. There's nothing pure, cold, and timeless about the future. Nothing at all. The world and its people are very provisional. Only their absence is permanent.

As you may have gathered, like a lot of authors, I'm a melancholic. You might not guess it to see me chuckling as I type this—for I consider this to be an amusing book, deliberately sunny and forward-looking. Still, all too often, even in these pages, I find myself having a whale of a good time savoring events and possibilities that are, by any conventional measure, quite grotesque or sad. Even though I've been favored by fortune, my imagination has an irretrievably cyberpunk, Gothic, matte-black-and-chrome cast. I am temperamentally attracted to dark, cranky, marginal figures.

But not to Dr. Pangloss, because he's too shallow and bland. Nor to Cassandra, for she lacks a genuine interest in human survival. And not to the Insurance Salesman, for though he pays well, he's got only one agenda; a meeting with him is everyone's byword for boredom.

Though human attitudes toward our prospects may differ, whenever tomorrow becomes today, Pangloss, Cassandra, and the Salesman live there at the same time. Today is cheery for some, grim for many, and in some sense always business as usual.

Tomorrow is another day, so for the sake of *Tomorrow Now*, I've decided to introduce a new futurist character as our guide and inspiration in the pages that follow. He's a famous and accomplished time-traveler, dressed for the long term, for he wears dark, hipster silks and satins equally suitable for the court, the jail, or the wilderness. He is "the melancholy Jaques" from Shakespeare's *As You Like It*.

The word "tomorrow" appears in *As You Like It* thirteen times. It's a great play for futurists becomes it smoothly combines contradictions. It's a cheery "comedy" in which all the good-guy characters are grim dissidents in political exile. The status quo is changing quickly and violently, because the central authority figure, Duke Frederick, is a paranoid lunatic.

One by one, the characters flake out and defect from the duke's cramped version of reality. Some are kicked out; others just scram; sometimes they run off in teams. They flee to the Forest of Arden, where they hope for better while living on venison, rather like Robin Hood. They all claim that they really enjoy this highly unstable situation, even though it's dirty and freezing.

Except, that is, for the melancholy Jaques. Jaques doesn't

have to pretend to like this uprooted life of exile and uncertainty. Because he *genuinely likes it.* A consummate observer, he feels quite at home at the edge of futurity. The other characters are stagily getting close to Nature and pretending to find "sermons in stones," but Jaques is the guy who is *putting the sermons in there.*

Despite the fact that he's a black-clad aristocrat, the melancholy Jaques has a contradictory hankering to be a colorful court jester. "Invest me in my motley," demands Jaques,

> give me leave
> *To speak my mind, and I will through and through*
> *Cleanse the foul body of th' infected world,*
> *If they will patiently receive my medicine.*

Spoken like a true subversive. For all his arty theorizing, this guy wants real-world results.

Eventually Duke Frederick has a major mental breakdown. His time is up and his regime collapses. Somehow a pagan goddess arrives, and everybody gets married. Then they all scrub off the dirt of the wilderness and head for the cozy certainties of their restored status quo.

Except, that is, for Jaques. Jaques is *staying out there.* At the play's end, he prophesies happy endings for the other folks, literally brushing them off the stage. But Jaques himself isn't leaving the Forest of Arden. Instead, he placidly hangs around to debrief the insane duke. He has no reason to leave his permanent uncertainty. For him, it's the very best dwelling place.

Jaques is best known for his famous set-piece lecture. This is his telling disquisition on the stark relationship of time and human beings. It goes like this:

> *All the world's a stage,*
> *And all the men and women merely players;*
> *They have their exits and their entrances,*
> *And one man in his time plays many parts,*
> *His acts being seven ages. At first, the infant,*
> *Mewling and puking in the nurse's arms.*
> *And then the whining school-boy, with his satchel,*
> *And shining morning face, creeping like snail*
> *Unwillingly to school. And then the lover,*
> *Sighing like furnace, with a woeful ballad*
> *Made to his mistress' eyebrow. Then a soldier,*
> *Full of strange oaths, and bearded like the pard,*
> *Jealous in honour, sudden and quick in quarrel,*
> *Seeking the bubble reputation*
> *Even in the cannon's mouth. And then the justice,*
> *In fair round belly with good capon lin'd,*
> *With eyes severe, and beard of formal cut,*
> *Full of wise saws and modern instances;*
> *And so he plays his part. The sixth age shifts*
> *Into the lean and slipper'd pantaloon*

With spectacles on nose and pouch on side;
His youthful hose, well sav'd, a world too wide
For his shrunk shank, and his big manly voice,
Turning again toward childish treble, pipes
And whistles in his sound. Last scene of all,
That ends this strange eventful history,
Is second childishness and mere oblivion,
Sans teeth, sans eyes, sans taste, sans everything.

(*As You Like It*, 2.7.147–174)

This is the future as it is felt and understood: via human experience.

My intent in writing this book is to come to terms with seven novel aspects of the twenty-first century, situations that are native to that epoch and no other. These are seven challenges, driving forces, opportunities this century possesses that make it distinct from the past. I'm presenting these future possibilities in the way that people experience time: in thematic stages.

Stage One: The Infant concerns genetics, reproduction, and microbiology. Stage Two: The Student is about information networks and new paradigms for the scholar. Stage Three: The Lover takes its cue from *Pygmalion;* it is about postindustrial design and our fiercely passionate relationship to our own creations. Stage Four: The Soldier is a war story about the growing New World Disorder, the new century's greatest security threat. Stage Five: The Justice tackles media and politics. Stage Six: The Pantaloon is a primer on twenty-first-century information economics. Stage Seven: Mere Oblivion is about our struggle with mortality and our technical assault on human limits.

The years to come are not merely imaginary. They are history that hasn't happened yet. People will be born into these coming years, grow to maturity in them, struggle with their issues, personify those years, and bear them in their flesh.

The future will be lived.

STAGE 1 THE INFANT

And one man in his time plays many parts,
His acts being seven ages. At first the infant,
Mewling and puking in the nurse's arms.

The infant personifies the future. You place your children into history. You are their past.

Futurists like to study population growth and trends in demographics, which is to say, people having children. The infant is no mathematical abstraction, though; a baby is the future howling aloud. Tomorrow now, born naked.

The delivery room is a place of primal hope and fear. It's a dramatic arena of suffering and risk. Few things are as common as a human child born all right, but when the futurist's own child is the hostage to fortune, there are very few comforts found in statistics. What if the baby dies? What if the mother dies? What if the baby is born deformed, with decades of sorrow ahead? The clock ticks, a child comes into the world, and no amount of rational analysis will stop that process. People must live with the consequences—because people *are* the consequences.

I like to think that as a father-to-be I fully deserved my many anxieties. Childbirth was certainly the most profound encounter with the future I have ever had. But unlike millions of jittery fathers in the past, I had a benefit in my possession that lacked historical precedent. I had a pocket photo of my child, taken before she was born.

I had a sonogram. It was a printout from a medical scanner. Its sonar nozzle had slid all over my wife's distended midriff, greased with clean medical jelly. The doctor had to wiggle this device about a bit, and peer and head-scratch through its Delphic, futuristic blurring, but he did it in real time and right in front of us. The child's limbs were in order, the growth numbers looked right, and to judge by the sonar shadows of her little pelvis, she was a girl.

What comfort we took from that technological artifact. With a sonogram at hand, you can abandon half the book of baby names. You can spin new plans for the colors of the curtains and the bassinet. This sonogram was like prenatal radar, full of swimming promise. Primeval darkness had left the womb. Its silent inhabitant was no longer a "pregnancy." "It" became "her."

That is how I first glimpsed my daughter: through an instrument. But my daughter did not, in fact, begin as an infant, or even as a sonogram. She began, just like her dear mom and dad, just like you, as an anonymous entity the size of a pencil dot. Humanity's origin is in the realm of the microscopic. That is the true start of our story.

Human eggs are minuscule, but we moderns can see them. They're no longer metaphysical, they're not folk legend or fertility ritual. They have become the province of rapidly advancing biotechnology. Single cells can be measured and manipulated, extracted and preserved. What we can see, we can sort, shape, and sell. We penetrated the realm of the microscopic with ever-growing technical sophistication. In the twentieth century we came to realize, with growing excitement, that the general business of life on Earth all runs on the same hardware. It's all cells, and at the centers of cells, it's always DNA. The business of life is Life-on-Earth Incorporated and Unlimited, a wholly owned subsidiary of deoxyribonucleic acid.

Genetic engineering is the twenty-first century's own new baby. In the century's dawn, biotech is its star turn. Biotech is by no means tomorrow's only major technology. The twenty-first century has the whole technological family crammed under its roof, fork in hand at the trestle table, a vast clan of hungry transformations, many of them centuries old: printing, clocks, railroads, electric power, radio, television, air flight, nuclear fission, satellites, and computation; it has the works. It's an orgy of sibling rivalry. But genetic engineering is tomorrow's native-born contribution to that family. It's the newest, the riskiest, and if it survives and flourishes, it will become the most powerful. Biotech is a baby Hercules that wants to kick the slats out of the crib.

Babies don't stay babies. My first daughter, for instance, is for the moment a thriving teen. Her rocketing passage toward maturity is written all over her; every day sees her blatantly learning and growing. Biotech is the baby industry now, but when it's big, it will reshape reality. To describe a biotech world, a world with a mature genetic technology, requires a new language. A new vocabulary, a new set of assumptions, a new literacy.

A baby, once she gets going, does not stop. It's a very different world, the future, but we're never going to "get there." There's no place "there" for us to get. The future is a process, not a theme park. The future itself *has* a future. We, in this present moment, are part of the future's past. The future is not an alien world, it is this very world, with different people, at a different time. Yesterday, today, or tomorrow, the clock never stops ticking. Every new stage must grow on the mulch of the last.

Bearing that in mind, let me introduce you into a biotech world. Here you are, let us say, reading a book. Not this book (unless you're some kind of antiquarian) but a similar one. Are there books in your biotech world? Yes. Made of paper? Sort of. Is that ink? Not ink as ink was previously understood, no; but why would you bother to notice that?

Let me make a few impolite personal observations as you sit there reading. By twentieth-century standards, you don't look very clean. In fact, you look rather greasy, and you're somewhat odd-smelling. But you are impressively robust and glittery-eyed, and full of animal vitality. Even though you are a harmless reader of late-twenty-first-century pop-science books, praiseworthily engaged in the intellectual trends of your own decade, you don't look especially scholarly. On the contrary: basically, you look like an athlete or supermodel. You look that way not because you're all egotistically eager to stand out from the norm but because that *is* your norm. An athlete or a supermodel is what men and women are willing to pay to look like. In your epoch, flesh and the processes of its construction are very ductile. You have no tooth decay, no dandruff, no enlarged pores. Though you read too much, you have no glasses.

Your home is snug and elegant. Its walls, floors, and furnishings are made of warm, organic substances that resemble cork, bamboo, and redwood, although they aren't. The lawn outside your membrane window has eight or nine hundred species living in it. It is a biodiverse menagerie.

You're just a normal person in a biotech world. You are not some grand chrome-dome master of biotech—no single mind can ever master such a broad field. Biotech is not even your personal line of work; you just live there. Your lawn is aswarm with living things because of social pressure from your neighbors. A mowed lawn is a scandal; you wouldn't subject the neighborhood to such a sight any more than you'd shave your children's heads to eradicate lice. You don't go out there and garden it, either. The lawn tools know more about plants than you do. And they work by themselves. It's a city lawn, not a wilderness. It's autogardening. The "wild" animals living in it don't know they are under surveillance.

Out on the street are scarab-colored nonpolluting vehicles that run on hydrogen. Like most industrial objects, they rot on command and return to harmless compost. Then there's your plumbing, or, as people put it nowadays, your "waterworks." In a biotech world, water networks are a bigger deal than bit streams. You're not made out of digital bits—like all living things, you are made mostly of water. So that's where you sensibly place your high-tech investments.

You don't have a "shower stall." You have a standard, everyday body-imaging system that gives you complete interior and exterior health scans every morning as it washes you. Your toothbrush scans the contents of your mouth and catalogs its microorganisms. Your toilet is the most sophisticated network peripheral in the home. It provides you with vital metabolic information about your body—the substances that enter and leave it and the vital processes within it. Only fools are squeamish about this.

Your bathroom cabinet is full of unguents, greases, and perfumes. There are some pills in there, but most of them do not contain drugs. Instead, they contain living, domesticated organisms that *make* drugs while living inside you. Some of the "pills" are cameras, with tiny sensors and on-board processing. Nothing in your medicine cabinet is sterile, not even the bandages. Modern bandages contain living organisms that are *good* for wounds.

"Sterility" is what people do need when they don't know what's happening on a microbial level. In a biotech world, sterility is a confession of ignorance. It's a tactic of desperation.

In your kitchen, the mops have more processing power than twentieth-century national bureaucracies. Your kitchen is mostly a place of filters and membranes and films; it is certainly not a butcher shop or a place to process raw vegetable matter. You eat delicate and tasty knickknacks that differ radically from grotesque historical foodstuffs. You have no fridge, because nothing in your house ever rots without your permission.

Even though this is a genetically altered world, there are no weird-looking "mutants" or "monsters" in your house, neighborhood, or city. You don't have, for instance, a six-legged dog. The cop on the beat is not ten feet high, and she does not look like RoboCop; if she has a baton, it doubles as a swab. It's not that such things are impossible for you and yours. Of course they are possible, but they are also crude publicity stunts dating from the eye-goggling infancy of biotech. In a mature biotech world, such nine-day wonders are considered crass and corny. They make no common sense.

Back in the early days of harnessing DNA, people were always fussing about full-grown multicellular beings—genetically altered humans, plants, or animals. There was a lot of anxious talk about clones (genetic duplicates) or chimeras (creatures with fused cells, whose bodies are mosaics of different species). Genetically modified organisms contained snippets of alien DNA, such as the artist Eduardo Kac's rabbit "Alba." That arty little rabbit, infused with jellyfish genes, could glow bright green in public. Alba the rabbit made a well-nigh perfect art-world cause célèbre at the dawn of the twenty-first century. Alba really panicked the bourgeoisie and was a nice *succès de scandale,* a worthy credit to the social insight of the artist. But once you'd manufactured a glowing green rabbit and shown it off, why would you ever want or need more than one?

For you, a modern DNA-literate person, weird animals have very little to do with the actual, real-world genetic industry. Frankly, the flesh of full-grown plants and animals just *gets in the way.* They might be dramatic examples of the concept (the way humanoid robots were once dramatic versions of the concept of automation). But the "threat" of "automation" turned out to be mostly hokum, and there were never any humanoid robots clanking around in real life, working on assembly lines. The same objection goes for monster Frankenstein animals. Yes, they sound really cool and scary, but go ahead, make one. Where is the market?

Expressing DNA in the genomes of large organisms is slow and clumsy. Creating an animal means deputizing some large and reluctant multicellular bureaucracy to carry out your will. That is not where the action is. It doesn't take efficient, industrial advantage of the raw power of DNA as a means of production.

Livestock requires long, solemn months of growth and delivery, just like a human baby. That is not industry, that is traditional unskilled labor.

All the real DNA action is in single cells. A genuine genetic engineer cuts to the chase and ratchets right down to the molecular hardware of the famous double helix. This news is no great surprise to you, for you were taught all this in grade school. You were shown the proper instruments for the job. You got down to the microbial level where DNA does all its heavy lifting, and you *stayed* down there. You marinated yourself in that seething point of view. You got all cozy with it. You got used to it. And if you happen to work there—and most people of your epoch have at least some kind of brush with DNA, the way most people used to have a nodding acquaintance with computers, or cars—then you study it and record it, analyze it, sequence it, copy it, map it, tag it, recombine it, commercialize it, exploit it, buy and sell it every day. It's how you modern folk live.

You know that DNA is not just a big molecule. DNA is

history. Like a baby book, DNA is a personal archive, full of profound revelations about your identity. You shed clouds of your personal DNA wherever you go, the material evidence of your life and your flesh. DNA carried ethnicity out of the old-fashioned world of folktales and flag-waving and into your world of factual, measurable relationships between chains of human code. So DNA isn't "a molecule"—DNA is us.

Your ancestors knew just two kingdoms of earthly life: plants and animals. You know more than seventy. Most of those kingdoms—vast realms of metabolic activity—belong exclusively to the single-celled. That's where the variety is, where DNA's skill set has been best developed. Microorganisms were busily manipulating DNA for three billion years before anything multicellular showed up on the scene. Most life is, and always has been, microbial. The variety of microbes is colossal, much wider than that of all multicellular animals. There are microorganisms living in boiling water and eating cyanide and sulfur. They're as different from humans, and from one another, as tigers are from cabbages.

As a DNA-literate person at ease with these facts of real life, you know that genetics is not a realm of boffin technicians in white lab coats. White lab coats are absurd to you, hopelessly old-fashioned. Lab coats were designed to show spills, so that they could remain sterile. For you that garb is like the armor of a medieval knight. If you spill anything

remotely dangerous or bioactive on yourself, the doorway
will tell you; the bathroom will tell you; a taxi, an air condi-
tioner, a stove can tell you. A five-year-old child can tell you
not just that you have an influenza virus but what kind you
have and where it came from.

You're into germs. Oh sure, you've got a cat; people who
read books like cats. Sometimes they even like cat books. But
you'd never expect your cat to do any industrial heavy lifting. Besides, your cat doesn't live *inside of you.* In a biologically savvy world, *inside of you* is where it's at.

You're into germs because germs are into you. No man
ever walks alone. Every human adult carries about two
pounds of living bacteria, or about a hundred trillion nonhuman cells. This is entirely normal and good. It's something
you understand about the real world that twentieth-century
people did not see and could not perceive. They had this
crude, desperate insight they called "sanitation," while you
possess a genuine insight and a hands-on technical mastery
of that situation. Unlike those blind primitives, you walk
your seething Earth in an aware, fully engaged, progressive,
civilized fashion. You swarm inside and out with microbes,
and it's *good for you.* You recognize and celebrate this. People
chat about their germs over coffee—it's like comparing perfumes. In your world, germs *are* the perfumes. Anyone who
smells bad is an utter ignoramus.

Your mother gave you her mitochondria when she gave
you life. Mitochondria are formerly free-living organisms

that have been perking along in the cells of human flesh for several billion years. They seem genetically foreign, yet they are also a vital part of humanity. Without mitochondria, we have no energy. It follows that to be truly antiseptic is instantly fatal. To lose your mitochondria "infection" would mean to die horribly, reduced to a flaccid bag of jelly. For you, losing your favorite microbes in and on your skin, bowels, and organs would be a grave environmental setback, like losing topsoil and songbirds. As for your handy mitochondria, you're very interested in souping them up. They are your beloved little engines—and you want some *heavy-duty* ones.

You, the thoughtful citizen of a biotech century, find it quite easy to think this way, because it is necessary. It is a basic requirement of your times. It has changed your assumptions and your vocabulary; it has shaped your very character.

It would have seemed odd and far-fetched to your predecessors. You're aware of this, because you are civilized and literate and you enjoy reading, let us say, Robert Louis Stevenson. Stevenson was the literary scion of a race of great nineteenth-century engineers. He once thoughtfully wrote that there are certain regions of scale and size where there is "no habitable city for the mind of man."

But your own mind does inhabit that city. So you know that Robert Louis Stevenson is a mental antique. For you and the people of your historical period, germs are not at all mysterious or alien. Two hundred thousand of them could lurk under the top half of this semicolon; but for you, domesticating beasts three microns long is not a problem. You know your own insides like the palm of your hand, because for you, the hand and gut are both important parts of your body, almost equally easy to see. Germs cause you no fear, bewilderment, or disgust. You entertain no such primeval superstitions. You properly envision a germ as a powerful entity, deserving of your gratitude, respect, and mature caution—maybe something like a stud bull. Your body contains millions of microbes, altered to do your will.

If you choose to work directly in the industry, you might regard microbes less sentimentally, say, as a big, sloshing multiton railway tank car. Germs are shaped like tank cars because osmotic forces inside a bacterium can reach seventy pounds per square inch, five times atmospheric pressure. And when properly altered, germs can manufacture almost anything you need. A tree, for instance, makes lumber. But to do so, this multicellular life-form has to sink its roots in the soil, hold itself up in storms and winds, ship sap up and down, grow fruit and seeds, build and spread out its solar collectors. That's great for trees, but for you it's a huge amount of unnecessary industrial overhead. If you want a construction material as tough as wooden boards, you don't chop down trees: you train some germs to make cellulose in bulk.

When it is properly fed, a bacterial tank car can build an

entire duplicate tank car from scratch. Not in twenty years, like a pine tree—in *twenty minutes*. Rapid, geometrical reproduction is why bacteria are everywhere in the world. It's also why diseases can kill people: they kill with their vastly booming force of numbers. Bacteria possess the power of DNA at its rawest, fastest, cheapest, toughest, and most immediate.

A bacterium also happens to be a splendid genetic engineer. A eukaryotic cell—we humans begin as one single eukaryotic cell—possesses a nucleus of DNA that is firmly coated in a membrane shell. Our DNA is stodgy, fail-safe, uptight DNA, suitable for mammals. Bacteria, by stark contrast, are prokaryotic cells, the oldest known form of life. Their DNA simply sprawls out amid their cytoplasmic goop like snarled and knotted Slinkies.

Bacteria create, manufacture, swap, and share enormous amounts of DNA. Not only do they share DNA among members of their own species, through conjugation and transduction, but they will encode it and ship it to other species. They can find loose DNA lying around from the burst bodies of other bacteria, and they can eat that DNA like food and then make it work like information.

Bacteria also make and carry plasmids, which are little rings of spare DNA. These DNA cassettes have various handy everyday uses for the bug that owns them. And bacteria have been doing this for billions of years. They're quite good at it. Genetic engineering is their way of life.

This surprising and fetid gene-splicing orgy isn't what Darwin imagined when writing *Origin of Species* in 1859. Gregor Mendel did not have this in mind when he was discovering the roots of classical genetic inheritance in peas back in 1865. But bacteria aren't peas. They don't work like peas, and they never have. From a DNA-industrial point of view, bacteria work much faster and much better than multicellular creatures. Bacteria do extremely strange and highly inventive things with DNA. And if we master them, then so can we.

In your world, bacteria practice birth control. Therefore, they are not contagious; they are little chemical factories. Unlike conventional livestock, bacteria never grow old or wear out. Better yet, thanks to a handy human-invented technique called "bacterial artificial chromosomes," it's rather easy to jam alien DNA inside bacteria and have the bug put that DNA to work. Sterile bugs sit still for your convenience, little tank cars there on their microscopic railroad siding, chugging indefinitely, spewing very useful and valuable compounds. They turn raw, cheap chemical feedstocks into almost anything that DNA can make: proteins, hormones, drugs, antibodies—and structural materials: skin, horn, bone, coral, bamboo, plastics even.

Furthermore, since germs are very low on the food chain, bacteria are also quite efficient at turning raw materials into substances that people can metabolize. In other words, food. The objection might be made that people don't much want to "eat germs." We have no choice in this matter. Germs already *eat for us.* They've been assisting our digestion since the dawn of time.

For you, germs are the world's major form of agriculture. Human pressure on the environment has dropped drastically. You've long since given up fussing over eating "genetically modified organisms." You're settled down to feast on genetic modification itself.

Genetics, you see, is a cultural point of view. It's not a matter of weird machines, the sequencers and chromatographs and so forth. Those mere devices quickly become obsolete. It's a matter of seeing the once-secret productive engines of nature, envisioning that as an industry, and understanding the broad implications. It is a worldly philosophy. It's like seeing a grainy shadow on a sonogram and understanding its potential as a little girl.

Even in your (imaginary) world of a mature genetics, a late-twenty-first-century world where genetics climbed out of its crib and became a major world-altering captain of industry, your life is not utopian. You're not some ultimate master of DNA power fantasies. Biotech and its products are around you every day, but you don't bone up on the subject every day, any more than everyone in the rail-crazy nineteenth century worked on a railroad. Biotech is what is modern. Your ability to adapt to modernity is a matter of will and circumstance, spiced with catastrophe.

If genetics becomes a major industry, then like the others

before it, it will be extremely complicated, with avant-gardes and rear guards, fads and fallacies, boom times and crashes. It will be a major long-term technosocial transition that is hard and multiplex, full of contradictions and backwaters and stubborn silences.

It will also be irreducibly controversial, even when it's decades old and very advanced. Darwin's *Origin of Species* is the inspiration of genetics. Knowing that creatures inherit, mutate, and evolve: this was the conceptual key without which genetics would make no sense. Darwin's book was written way back in 1859. In 2059 people will still be in frank denial about its revelations. They'll even be in open revolt. They can't bring themselves to see it; evolution is just too much for them to get their heads around.

Biotech is different in character from its older relatives, those major, visible, clanking industries. It's not a stiff, armored technology like nuclear power, which is large, remote, and cyclopean. Nor is it like spaceflight—glamorous, distant events occurring in orbit under the control of armies of experts. Genetics is intimate—every bit as intimate as any technology can get. It is about personal identity and heritage, blood, bone, egg, and seed. It cannot be dealt with as the nineteenth and twentieth centuries managed their machineries. Biotech industrializes life, our very substance. It's best compared to organic phenomena: fertility, yeastiness, and contagion. Biotech doesn't clank and beep. It seethes and bubbles and leaks.

People in the twentieth century made a sharp distinction between the abstruse, tiny realm of molecular chemistry and actual people walking around loose in daylight. But tomorrow that pretense will hold no water. Because it's all the same.

There is always a lag between what we know and how we feel about it, between scientific data and its social meaning. We'll know we've made real progress when we've come to new verbal terms with sex. Sex is a very likely arena for a new, popular genetic vocabulary to be invented, because sex is very slangy, and sex is also human DNA in action. In sexual matters the twentieth century departed radically from the hushed Victorianism of the nineteenth. The twenty-first will likely do the same for the twentieth, abandoning the lenses of psychoanalysis and sexual liberation to recast the issue in some new native jargon of biohardware and hormonal software.

There are excellent reasons to do this. At the turn of the twenty-first century, a chronic disability to talk about bodily fluids is killing huge numbers of people. Sexual reticence has social and medical consequences severe enough to depopulate Africa.

The lack of common definitions leaves the social landscape mired with paradoxes—the traps of of ideological struggle. Americans cannot agree whether a microscopic human entity is "alive." So in America "abortion" is both a "civil right" and "murder." Human stem cells hover in a paradoxical twilight zone, somewhere between personhood and patentable hardware.

We have no common language to describe what goes on in the womb. It's not mere privacy that makes us balk and stammer; in the past, we simply didn't know. What little we did know was very badly phrased. There's nothing life-affirming or attractive about the stark Latinate terms that describe human conception, such as "corona radiata" and "cumulus oophorus." That terminology is dead on arrival.

Since conception is sexual, it is probably best understood and described as something really sexy. Conception may be shrouded deep in our flesh and distant to our everyday experience; but we can see it now, and it is not a medical abstraction. It is extremely human. It may not *seem* very sexy—like sexual intercourse, it's one of those beautiful, meaningful acts that always seems really dirty—but it surely has strong erotic elements. The conception of a new human being is probably the most erotic event that can happen. It means a fresh entity emerging into the virile, fecund world of flesh and sensation. There's a human need to celebrate this; it should be a source of profound joy to us.

A term like "mitochondrion" has a formaldehyde tang. It sounds like an itch or a disease, rather than the most genuine and primal source of all human energy. We lack a positive way to describe a joyful and life-enhancing infestation of our flesh by tiny microbes. But we rather need a word for this, because the human mouth alone shelters some 450 native species. When we can talk freely about this reality and enjoy it for what it is, we will have come to proper terms with biotechnology. But alas, the century is very young, and that sense of mature serenity is still some stiff distance off.

Genetic engineering may someday become a mature industry, even old hat—but it will never be perfect or magical. It may refine its jargon, but it will never be entirely well controlled. If there is any activity that DNA truly excels at, it is proliferation. Multiplying and spreading is DNA's very reason for being.

Genetics is a true frontier today. Its pioneers are blundering around looking for some way, any way, to make a profit, and to finance further research and development. They do some very cruel and clumsy things: the period equivalent of sacking and looting the New World in a search for gold.

Consider the wacky and distasteful practice of embedding poisons into foodstuffs—crop seeds with insecticide genetically built in. This might make sense from the point of view of shelf space at the farmer's supply store, but in terms of a rational course of industrial development, it's extremely shortsighted. It is also politically disastrous. People are naturally and reasonably sentimental about large, cuddly organisms such as wheat and sheep. Sheep are in children's nursery rhymes. Amber waves of grain get poetic mention in national folk songs. Messing about with wheat means directly tampering with apple pie. Only fools rush in here.

From an aesthetic point of view, it's hard not to flinch at the crude violence done when tomatoes are infested with fish genes. This kind of thumb-fingered intrusion is the tell-tale sign of an immature technology. In retrospect, such acts will be regretted.

Bacteria, on the other hand, have no sentimental cultural defenders. Better yet, they are already accomplished genetic engineers. We'd be wise to thoroughly study their work before firing our own commercial shotguns in the dark. We don't merely genetically work on bacteria; they have genetically worked on us. Through an eerie process daintily called BVTs, or "bacterial-to-vertebrate transfers," bacteria seem to have stuffed quite a lot of their DNA into the human gene pool. By one recent speculation, some 233 genes in the human genome may have been crammed into us by "lateral transfers" from bacteria.

Since they've had millions of years to work inside us, it's plausible that bacteria have, on occasion, penetrated human egg cells. A germ called Wolbachia has done quite astounding things to sexual reproduction in insects. Are interested microbial bystanders really up to something peculiar in the human egg and sperm? Who knows? In our arrogance, we've never thought to look.

The twenty-first century brought something new to the ancient world of the microbe *Escherichia coli*. Intelligent human beings raided *E. coli*'s genetic library and copied the works. All of the DNA in *Escherichia coli* has been fully sequenced: its 4,600,000 base pairs and 4,000 genes have been sampled, named, and numbered. In September 1997 they were published.

The same event occurred to us humans. In February 2001 the International Human Genome Sequencing Consortium published an extensive full-color map of humanity's genetic workings. Homo sapiens has joined the ranks of the genetically sequenced, along with thirty-one bacteria, a fungus, two animals, and a plant—with more candidates for scanning almost every week, including even the Black Death.

What happens after that knowledge? Well, to see it means to change it. "Genetic engineering" is generally thought to mean changing DNA in some brisk, direct fashion, by complex technical means. But as our friends the microbes have demonstrated, there are plenty of quick, effective, untechnical ways to change DNA. Genetic researchers can do a great many useful and profitable things to DNA without any genetic engineering.

Let's have a close look at the loopholes here. Suppose that "recombinant DNA" is—somehow—made illegal worldwide. Let's further suppose that you are a reasonably clever, law-abiding person who still wants to work in the genetics industry. Splicing together DNA from different sources into a single creature is now forbidden to you. For your genetic industry, this is a speed bump but not a road-block. You don't have to "add" any alien genetics to an organism. With "knockout" techniques, you can *remove* chunks of the organism's DNA and learn a great deal.

The vast majority of DNA appears to be archival, in other words, historical "junk." So you don't have to invent or insert any weird or foreign DNA in order to create novel organisms. Just recycle the junkyard! There are spare parts galore inside every organism. Junk DNA can be repurposed and reactivated. Heaven only knows what a pig has in storage in its ancient genetic attic. There used to be Paleolithic pigs the size of Volkswagens.

Traditional livestock breeding already combines DNA, through sex. If you test for desired DNA in the offspring, you can vastly increase the genetic efficiency of traditional sex. You're no longer guessing about results when you breed an animal. You hit the bull's-eye every time.

It takes a long time to breed a champion bull. But microbes can reproduce every twenty minutes. That makes seventy-two generations of breeding selection, every working day. Here you have a road wide open to genetic revolution. Just make the invisible visible; the rest will follow. There's no crying need for you to become Dr. Frankenstein. You need not rub your latex-covered hands together as Mother Nature quails in horror and the lightning flashes out your garret window. Just take a hard, fresh, serious look at the natural world. Look deep inside what you already have, what you already are. Come fully to terms with that knowledge. Knowledge is power. Options will multiply.

Technology never leaps smoothly from height to height of achievement; that's just technohype, it's for the rubes. In the real world, technology ducks, dodges, and limps. If genetics can't march to victory by the straight path, it can reclassify itself as "chemistry," or "medicine," or "farming," or even "nanotechnology." Then it can pursue the same goals under a different set of regulators.

Genetic industry is a newborn infant. Barely christened, fawned over by anxious venture capitalists and medical ethicists, it's mewling there in its crib, trying to find its toes. When it gets bigger, genetic engineering will radically expand our knowledge—especially our medical knowledge. However, it won't look or sound medical. It will break medicine open at the seams. It will reinvent its language and repurpose all its tools. The art of medicine is thousands of years old. It still speaks Greek and Latin. Its practitioners are bound by oaths that are older than Christianity. Its values are centered in ancient standards of sickness and health, not gene-centered values of implants, copies, and upgrades.

Medicine is concerned with restoring sick people to a standard of human health. Genetic engineering transcends this. It redefines health on a sliding scale, as an industrial process. A doctor asks you how you feel; a medical geneticist would properly ask how you *want* to feel.

At its core, mastering genetics means giving human beings some of those wonderfully elastic and vital qualities that formerly belonged to bacteria. If we humans ever become as skillful as bacteria, then our genetic engineering will resemble theirs. It will be common, cheap, temporary, and reversible. Alien DNA will plug in and out of our human flesh, and it will leave no trace when deinstalled. Our flesh will grow and decline on command, and it will be ageless. Our genetics will be scalable and upgradable. (It may be too much to ask for genetics to be virusproof.)

We can aspire to mimic the skill of bugs; that doesn't mean we'll achieve it. We can describe a geneticized world that seems safe, efficient, and logical—a world that fulfills the promise that genetics offers. But trend is not destiny. Even the best technical prognosis is merely a child-rearing guidebook. No kid in the real world ever gets raised by a book.

As the twenty-first century dawns, people are unhealthily obsessed with cloning human babies. This endeavor is certainly not the summit of genetic skill. It is one rather simpleminded stunt in a vast spectrum of genetic possibility. Every modern political authority frowns on this practice. However, it is technically feasible and increasingly affordable. There is a definite groundswell of interest in cloning people. It is propaganda of the deed.

Cloning human babies does not have much of a future. This may seem like a counterintuitive thing to say, but it quickly becomes much more plausible when we stop thinking about cloned babies and start thinking about ourselves.

Let's move from the abstract to the specific. Forget the

cloned baby. Instead, let's talk about you. *You* are the cloned baby. You are the child who has the rather severe social misfortune of being a clone. Your "parents," or rather, your technical sponsors, carried out their weird goal, and you were their end product. You are the child they have on their hands—not the child of their love, of course, but the child of their boundary-shattering technological ambition, carried out in the teeth of stiff social resentment.

Once upon a time you were a highly publicized, headline-grabbing baby. But you didn't stay a baby. Nobody ever does. Secondhand shops are full of baby clothes, scarcely worn. Babies always get bigger. Babies rise right up. They want to walk among us. This has been the baby chapter of this book: stage one in the Seven Ages of Man. This infancy section will be over very soon. There are just a few pages left! You're holding this very book in your hand—infancy is fading fast, but look how many ominous pages still lie ahead of you!

So never mind your bronze baby shoes and the shocked media coverage you once got in your bassinet. Unlike, say, a complaisant and brainless *E. coli* cell, you are fully capable of holding lasting grudges. And you surely deserve to. When were *you* let in on this cloning deal? Who asked *your* permission to transform you into a cutting-edge freak, a crass industrial experiment? You never asked to be born (or, in your case, cloned). You are furiously bitter and angry about this amoral imposition on your autonomous individuality. What are you supposed to tell your teachers about this? Your boyfriend? Your husband? Your employers? Your own kids?

It gets worse. Your personally lousy situation and your embittered prospects are not the full, grand story of human eugenic ambition. Alas, you are merely a passing stunt. We can forget about you; like the formerly marvelous test-tube child, you are so over.

There's little reason to create a cloned child when normal children are so available. The grander scheme in messing with the human genome is a genetic *improvement* of the subject: a superbaby.

Cloning might create genetic copies of, say, prominent politicians or pop stars. Given their difficult personal circumstances, these belated twins aren't likely to do well. However! Once it's possible to implant chosen genetics into a human egg and a womb at will, this creates an exciting potential for brand-new genetic installations. These people are a *new kind* of people—radical improvements on the sordid norm. Genetically modified superbabies surely seem much sexier and more dramatic than mere clone babies.

Learning nothing from experience or common sense, our malefactors boldly advance into the superbaby business. Their aim is a fabulous new-model human with—just imagine it—Olympic strength, maybe, genius IQ, drip-dry hair, Teflon skin.

But never mind the stellar publicity bonus. Technologies limp. Technologies never make a sudden Nijinsky leap from here to Oz. First, logically and necessarily, must come the *alpha-rollout* of a superbaby. Not some ultimate Nietzschean superbaby, mind you, but first, the *experimental* superbaby. The *guinea pig* superbaby.

And this superbaby is no mere abstraction; this baby is, once again, you. You were that infant in question, and yes, by human standards, you were indeed a superbaby. But never mind those humans. Their standards can't apply to you. Except for your moronic "parents," humans offer no standards by which you can be judged. Since you are "super," you are painfully subjected to an entirely *different* set of standards.

By generic superbaby standards, you are, once again, nothing much. In fact, you are lousy, and you know it. Why? Because the first superbaby cannot be the best superbaby. How could *you* possibly be the best? Nobody's had any real-world practice at creating ultrapeople like you. You are a beta-release superhuman. You are merely a prototype. And since the people who built you are incapable of thinking through the issues, you are almost certainly a hack job.

Foolish people yielded to temptation and created you, because that prospect sounded exciting to them. But now that you have reached majority, you're not particularly "super"—you're a prototype with a raw deal. Why? Because of the nature of technological advance. When you finally become twenty-one years old, genetic understanding is twenty-one years deeper and broader than it was when you were

first put into production. So you're not only a hack job—
you're also an antique.

Two decades is a very long time in a technorevolution.
Two decades is the distance between the technologies of the
First and Second World Wars, the difference between the
lumbering Sopwith Camel biplane and a supersonic V-2
rocket. You finally reached young adulthood only to find
yourself obsolete. There are already better, faster, and
cheaper ways of doing whatever it is that you were genetically altered to do. And these alterations are probably *not*
inscribed within human eggs, the way your alterations were.
That is a hopelessly slow and clumsy way to handle the
power of genetics. You are stuck with hardwired genetics
inside your own flesh. "Normal" humans *swallow* their
advanced genetics, genetics that they can plug and play.

Your contemporaries use their modern microbes to do
whatever your "supergenes" are supposed to be doing for
you. They never had to undergo the shattering trauma of
being superbabies. They sensibly let microbes undergo all
those hazards. Their microbial bugs may go obsolete, but
those are replaced on a technical whim. Microbes have no
problem with becoming obsolete; they don't mind it a bit.
You, however, are becoming obsolete, and you mind it
plenty. You, the superbaby, once looked like the greatest
baby in the world, but after your babyhood, you really got
burned by this deal. You have every reason in the world to
sue. You should supersue.

The outcomes of baby engineering are heavily weighted toward the tragic and the shameful. The risks, even in a technical "success," are very great. You, the beta-release superbaby, were cruelly manipulated without your own consent. You were brought all unwilling into a world you never made. You were an uninformed test subject, an abused child deprived of human rights. Your own creators—your so-called parents—are clearly your worst enemies. If you are bright, you will be bright in doing them harm. If you are strong, you will be strong for their evil. If you are eloquent, you will be eloquent in reproach.

So, even if we crassly set all ethical issues aside, the benefits of perpetrating a baby stunt are doubtful. There are powerful reasons to do this, and it may well get done; but the likeliest candidates at the moment are crazy religious cults.

Corny ideas about the *uebermensch* are not native twenty-first-century ideas. They are outdated Victorian and Nazi kitsch from the nineteenth and twentieth centuries. They are not realistic concepts about the true potential of DNA; they are phony ideas about racial destiny. A racially pure eugenic regime, the Nazi Reich, was supposed to last for a thousand years. A thousand years is ten long centuries. No technology, and certainly no technological society, ever lasts that long. Knowledge is not and never can be that static. A thousand-year regime is a mythological monster.

Messing with DNA won't and can't create Men Like Gods with a utopia in their tow. In taking command over

DNA, we are not taking command of our evolution and our destiny. We are merely gaining a transformative knowledge and multiplying our future options.

There are wave after wave of potential applications in genetics. Genetics is the primal bedrock of both medicine and agriculture. Farming and healing are two lines of human work that are prehistoric, planted at the very basis of civilization. During the whole course of history, healing and farming have been in continuous, roiling development, full of excitement and crises and massive, terrible failures, with nothing less at stake than our life and our death. Genetics cannot make either of these ancient arts more stable. It will drive them both into stampede.

As this first chapter ends, we can sum up infancy in this way. When your own child is born, it's a very big deal. It is a concrete demonstration of your adulthood, your fertility, and your potency. Yes, it's quite the step up, in many ways. It's rather gratifying to one's ego. It is maturing and quite educational. It radically expands your outlook on human life and its generational continuities. Reproduction has a lot to recommend it.

However, it is anything but simple, clean, minimalist, and utopian. Infancy is the last place to look for stasis or perfection. Childbirth is the polar opposite of all things remote, detached, and sterile. Childbirth is the future, the genuine version. Having a baby around is a wailing, sleepless, septic mess.

That is what tomorrow will look and feel like with a lot of seething DNA around. Genetics means new life, reproduction, with all its grave hazards and without any guarantees. Like the birth of a child, it means furious arguments, grim responsibility, anxious compromise, and seriously wrecked routines. It means ear infections, weird stinks, constant nervous monitoring, medical checkups, and postnatal depression. It means that we're always watching, always engaged, always worried, always tied by the heartstrings, and we never get off that hook of responsibility, not until our children bury us and we become the dead past.

We are the raw material. Biotech is us, industrialized.

Technology always "improves," but the wisest path forward is a path that allows us to keep making fresh mistakes. When we're dealing with genetics, the stuff of life, we need to shy strongly away from approaches that are irreparable and can work us into a fatal corner: monocultures, monopolies, and the obliteration of alternatives. We mustn't fly before we can walk.

Everything we do with genetics in the next few decades will look very primitive, even stupid and counterproductive, in the full light of a more sophisticated understanding. We should never design anything genetic without an "undo" button handy. We should not make permanent decisions in a state of mind that can't last.

Let me offer a last and quite cogent consideration. There will always be large numbers of our fellow citizens who refuse the language of genetics, who are implacably and effectively opposed to its view of the world. They will not go away, and they cannot be converted or won over. Anyone who expects smooth sailing for genetics when evolution and abortion are so ragingly controversial—well, that defines a fool's paradise.

It's a lot to ask that a technology should grow up under these trying circumstances. However, when you think about it, this is the human condition. People do grow up under such paradoxical circumstances. Genetics is the twenty-first century's new baby. So give the little kid a sandbox first. Don't expect him to play God.

A baby is small, but a baby is serious business. Confronting the future can mean genuine existential anguish. A baby can mean real, immediate, personal dread. There was a vital moment in my life when I knew full well that I might suddenly lose everything that I held most dear. A moment whose intimate consequences, all bitterly unpredictable, stretched not just through my own life but beyond my life into the lives of many others.

Then my child was born. I was there for the birth, in all its gooey, visceral glory. My daughter was placed in my arms. She was just fine. Toes, fingers, eyes, ears, a small but entire human being, a new Shakespearean player in the old human drama, shoved through the gate of futurity with a kick and a lusty wail. And I was really happy about that. I was overjoyed. I felt overwhelming relief and cosmic gratitude. It must have lasted an entire five minutes.

Then, of course, parenthood started.

STAGE 2 THE STUDENT

And then the whining school-boy, with his satchel,
And shining morning face, creeping like snail
Unwillingly to school.

When Shakespeare describes a schoolboy lugging his burden, these five-hundred-year-old lines require no footnotes. The ways in which we civilize children are very conservative indeed.

"Learning" is not the center of school life. Elementary schools are socializing institutions. They teach children to behave in civilized groups. As a kid, you don't get to go home when you learn the lessons of the day. If this were the case, all the bright kids would be gone inside an hour. Then they would devote their days to earnestly studying the complex topics that children find genuinely engaging, such as sports, cartoons, and video games.

No matter how clever they are, children are always kept in school till the bell rings. This teaches them to behave acceptably in large, bureaucratically organized institutions. They're also kept there in order to free up the productive time of their parents. Their parents, in theory at least, are already civilized. They supposedly work in situations that are even more stiff and constraining than those of their children.

At least, that was the case when public schools were invented. Personally, I work in the culture industry. As a novelist and journalist, I ply my trade mostly at home. My productive routine involves mushy, nebulous activities such as schmoozing with colleagues, websurfing, reading e-mail, avoiding conference invitations, and leafing through books and magazines. Sometimes I type. I do plenty of research, but my work has little to do with the rote labor typical of schoolwork. I have no rote and would be very alarmed if I did. I rarely fill out forms, and I never take tests or pop quizzes.

My older daughter, by contrast, is a student in high school. Compared with her lackadaisical father, she lives in harsh paramilitary constraint. She has a dress code. She fills out permission forms and tardy slips, stands in lines, eats in a vast barracks mess room. She comes and goes at the jangle of a bell, surrounded by hall monitors. She endures much paranoia about handguns and is frequently subjected to high-minded indoctrination against the illegal use of drugs and the careless exchange of bodily fluids. My child leads a narrow, tough, archaic working life. Though she isn't paid for her efforts, she'd do pretty well as a gung-ho forties-era Rosie the Riveter.

There's a lofty, patriotic element in American state education. So my daughter might also do very well as a New England cotton-mill girl in the heyday of Ralph Waldo Emerson. Back in those days, America was first making up its collective mind about the structure of the nation's intellectual life. A New England cotton mill was a factory, but not merely a factory. There was a lot of careful shepherding in those barracks, and anxious respectability, and much pious sermonizing that almost hid those racks of whirring spindles.

Today's schoolchildren are held to grueling nineteenth-century standards. Today's successful adults learn constantly, endlessly developing skills and moving from temporary phase to phase, much like preschoolers. Children are in training for stable roles in large, paternalistic bureaucracies. These enterprises no longer exist for their parents. Once they were everywhere, these classic gold-watch institutions: railroads; post offices; the old-school military; telephone, gas, and electrical utilities. Places where the competitive landscape was sluggish, where roles were well defined. The educated child became the loyal employee who could sit still, read, write, and add correctly—for thirty years.

Today's young students are being civilized for an older

civilization than our own. In hindsight, it's clear that my elementary school did everything it could to avert the career that I have today. My working routine is nothing like classwork. It's very much like the work of programmers and venture capitalists, those giddy myrmidons of a digitized economy. These people are intellectual entrepreneurs who have no institutional certainties and no well-defined titles or roles. They work, dress, and act like permanent grad students always denied tenure. As their contemporary, so do I. In the mid-twentieth-century Texas oil-company town where I grew up, it was never imagined that I would live like this and get away with it.

Thanks mostly to my traditional education, I used to think of my freelance writerly life as being very footloose and far out. After all, I'm a science fiction writer. It's a professional requirement to pride myself on being bizarre. Furthermore, I've been living through one of the fastest technological transformations in human history. It's a well-nigh permanent information revolution that claims to run twice as fast every eighteen months. Such was the core spiel of the Internet boom, which spewed more Technicolor visions than the psychedelics tent at Woodstock.

But in point of fact, I'm not very far out. Quite the opposite, really. I do entertain some wild ideas, but my personal life is predictable and tame. My dizzy, high-tech society is also far less frenetic than it looks. By historical standards,

I'm a remarkably conventional person, in remarkably stable circumstances.

My father and my two grandfathers considered themselves sensible, down-to-earth conservatives, but they had lives that were far more restless and chaotic than my own.

I hate to boast about my domestic tranquillity, since any random car crash, terrorist atrocity, or cancer diagnosis could ruin that for me. Still, one of the advantages of middle age is that the historical record speaks for itself. In the very midst of the fastest technotransformation in human history, I've had quite a tranquil time of life.

I've had the same career since I was twenty years old. I've been married to the same woman more than half my life. I've lived in the same town for twenty-seven years. I even have a thirteen-year-old e-mail address.

My father was born in a worldwide depression. He served in the Korean War. I never went hungry. I never spent a day in uniform. My grandfathers were Texas farmers. They lived close to the bone and knew real and backbreaking trouble from bank collapses, droughts, and bad crop prices. I've never watched my bank collapse. I've never been bankrupt (even though, as a freelance writer, I clearly have every excuse).

My life just doesn't much resemble the vivid, boisterous horror that historians have justly come to associate with most of the twentieth century. I've never been deported or sent to a camp. My hometown hasn't been shelled by enemies or torn up by race riots. Never once have I been shot, firebombed, or stabbed. That's what chaos and social breakdown mean, but to me those things sound exotic.

This is not to say that my life has been without giddiness or personal sorrow. It's by no means a tedious or sheltered life. What this does say is that America had a radical, across-the-board technological revolution, in social conditions of amazing peace and somnolence. The years 1989 to 2001 in particular were a belle époque. Those times might be best described as "edifying." That period's movers and shakers were a class unto themselves. They were hypereducated policy wonks closely focused on techno high performance.

Cyber high tech ran like crazy into every corner of society, and nothing awful happened. Nothing awful happened for an awfully long time. Even though swarms of high-tech machines were seething and breeding and obsolescing all around us. This just can't be coincidence.

It's no coincidence that my daughter is appalled by her schoolwork yet thrilled by the Internet. Loathing her official school assignments, she spends hours tracking down arcana on the Net, in patient orgies of pop-culture research. She hates the names and dates of dusty historical figures yet cheerfully memorizes the exotic dossiers of dozens of Pokémon. She dislikes geography classes yet thinks nothing of e-mail from Japan.

An information economy requires constant learning. That is why the wage value of a college degree increased radically during the tech boom, so that even liberal-arts majors often ended up in tech jobs. It wasn't the hardware that mattered but the ability to recomplicate information flows, to cut losses and grab the next new thing. By mastering new concepts and skill sets, people in and around high tech stay employable. They are also strictly required to do a great deal of forgetting—primarily about their industry's dead products, defunct companies, and severe financial busts.

In an information economy, prices and stock values are very volatile. Takeovers and acquisitions are endemic. Employee loyalties and management paternalism are archaic. Stock valuations bear little resemblance to the profits that a company can reasonably be expected to generate. These conditions are not regrettable accidents or oversights. They are an inherent part of the way an information society structures its civilization.

It's not that society has grown colder or more ruthless. It's that the doors and windows have been thrown open and the walls blown out, so that people, ideas, and money can dart in and out at the speed of light. An information economy is inherently low in backwaters, shelters, and sinecures. It lacks cozy places where people can potter along for decades, engaged in some single activity with some predictable rate of return.

It follows that, in an information society, a formal education aimed at vocational success would not be about values or canons. It would lack eternal verities, moral codes, constitutional continuity, literary classics, and good old-fashioned national heritage. It would lack the very things that teachers and scholars traditionally consider the sacred torch that must be passed to the coming generation.

An information society doesn't have time for cultural continuity and bedrock moral fervor. That is the kind of steadying education provided by fundamentalist Islamic *madrassas*. For a child of the Internet epoch, only a temporary canon would work, something hot-wired together out of spare hardware and multicultural scraps. Formal education for an information society would be about getting up to speed and staying there, with a minor in outside-the-box paradigm busting.

This imaginary form of education—the education fit for the future—would not talk much like governments, armies, or organized religion. It would probably talk in much the way that contemporary business magazines talk: with trend spotting, lots of spin, big money, and firmly crossed fingers. It would be a very faddish education. Large chunks of it would become obsolete in short order and collapse in embarrassing ways. People would be remarkably eager to unlearn things.

Academia in these circumstances would no longer resemble the nine-hundred-year tradition of the Western model. It would come to look more like industrial research and development: clever, fast-moving, but vapidly focused on products and profit. There would be no tenure, and few places for reservoirs of abstract knowledge to hide from market forces. Institutional behavior would likely change to mimic business practice; for instance, new academic disciplines might "acquire" old ones. Particle physics (corny and cold-war) might well be acquired by nanotechnology (sexy and booming). Fields of learning would ooze giddily from place to place, shedding any professor too stuffy to keep up: biology plus cybernetics becomes bioinformatics. Taxonomy minus natural history becomes genetic patent research.

Unfortunately, this speculative situation is not scholarship. Intellectually speaking, it means treading water. When you have no established canon of cultural classics, you have no place to take a permanent intellectual stand. You have no scholastic mastery, you merely have clever acts of opportunistic contingency. These losses are serious. Honestly confronting this stark realization leads to intellectual crisis. It causes "canon panic."

Canon panic is by no means a new thing, especially in America. The American intellectual establishment has always had an uneasy, catch-as-catch-can sense of permanent disequilibrium. Americans have always suspected themselves to be unhealthily obsessed with newfangled machines and money, while detached from eternal verities and culturally brain-dead.

In 1837 Ralph Waldo Emerson gave a famous speech called "The American Scholar." In delivering this manifesto, he sought to put a stamp of order on American intellectual life, to give it shape, substance, coherency, and a sense of direction. This noble activity separates a scholar from a rambling autodidact. A scholar possesses a framework of understanding. He doesn't merely surf about the intellectual landscape, picking up curios and assembling press clippings. A scholar has a calling, to organize knowledge in a verifiable, sensible, and schematic fashion. Scholarship makes humanity's knowledge conveyable and teachable.

Emerson knew the time was ripe for Americans to stop blundering about in sheer blithe ignorance and obtain an intellectual grip. "Perhaps the time is already come . . . when the sluggard intellect of this continent will look from under its iron lids, and fill the postponed expectation of the world with something better than the exertions of mechanical skill." In 1837 Americans were already world-famous for their habit of inventing and exporting weird gizmos. Emerson dared to hope for better. And who is to do this task for America, to free America from the bonds of technocentric cleverness, and to carry its people to the heights of a richer understanding? "The American Scholar!"

"The scholar of the first age received into him the world around; brooded thereon; gave it the new arrangement of his own mind, and uttered it again." Here Emerson urges us to make a higher sense of the hustle and bustle of existence.

"It came into him, life; it went out from him, truth." Now we're talking! You can sense those gathered Phi Beta Kappa scholars of 1837 leaning forward as Emerson clutches the wooden podium. "It came to him, short-lived actions; it went out from him, immortal thoughts." No mere sordid squabble for survival; we're composing the philosophical big picture. "It came to him, business; it went from him, poetry." America has always had plenty of business; if it were transmuted into poetry, there'd be stacks of poetry miles high.

Ralph Waldo Emerson, by dint of his heroic labor, did assemble a canon for the American scholar. It was called Transcendentalism. It's an eclectic, polyglot, rampaging New Age mess: it's Hegel, the Bhagavad Gita, Virgil, Homer, Thomas Carlyle, plus tips on worldly success and handy hints on how to merchandise better mousetraps. It comes from all over the planet; it's also peculiarly scattered in its erudition. But it couldn't be more American.

Emerson has enormous optimism. It never occurs to him to question this cheery feeling, and he just can't seem to help having it. It seems starkly detached from the objective realities of his situation. Emerson was a chest-thumping optimist in the days before the Civil War, which was certainly the worst domestic cataclysm that the United States ever suffered. But this looming horror on the future's horizon does not slow him down much. Fear of consequence never slows him down much. It's an aspect of national character that has persisted unchanged from his time to our own. Before the implosion of the dot-coms, the NASDAQ and New York Stock Exchange brimmed with irrational Emersonian exuberance. Millions of people cheerfully paying more for a software company than it can ever be worth—that is what Emersonian optimism looks like in a contemporary setting.

It is not very American to know how to make coherent intellectual sense. It's rare for Americans to have their children understand what they're talking about when they reminisce. But to be optimistic despite everything you've come to know about history is very American.

In particular, it is extremely American to get filthy rich doing something that nobody has ever heard of. Something that no authority figure ever gave you permission to do. In an industry that you could not be educated, trained, and licensed for, because it previously just didn't exist. An industry that is not only brand-new but that may, at any moment, vanish like a bubble.

Ralph Waldo Emerson had railroad networks and telegraph networks. We have information networks. But they are a revolution on much the same scale as Emerson's; they're just about as disruptive to our own preconceptions as Ralph's industrial revolution was for him. Railroads mingled goods and populations overnight, and telegraphs collapsed space and time. Our networks move almost everything on earth to a Web page three clicks away.

Contemporary society has invented a powerful teaching

machine that has no canon. This is the latest and most powerful collision of American gizmos and scholastic knowledge. Rather than being in harsh opposition, as they were in
Emerson's day, they have finally devoured each other.

The Internet has no curriculum, no moral values, and no
philosophy. It has no religion, ethnicity, or nationality. It just
brings on the data, railroad cars of it, data by the ton. The
Internet is scholarship that is electronically supercharged,
decentralized, disorganized, multicultural, and out of control. It runs on gizmos and it reeks of the transcendental.

Education is meant to codify culture. Computer networks cyberneticize this process, which is to say that they
give you everything all at once, from all over the world, at
the speed of light, in a harebrained heap.

No matter how vital American intellectual culture may
be nowadays, we're not likely to find another towering intellectual figure on the scale of Ralph Waldo Emerson. There is
just no elbow room for him. Today, armies of brainy pundits
and literati from scores of nations confront mountains of
fast-moving data. Emerson lived in a modest, barely industrial society where only a fraction of the population was literate. Where food and clothing were scarce. Where a sage
could talk to a roomful of college graduates and change the
course of a nation's intellectual life with one speech.

It might, however, still be possible to rise to bold Emersonian heights and assert some scholastic order on our multiplex life and our routineless routines. We can confidently

predict that should such a Yankee prophet exist, he would be oracular, opinionated, and senselessly optimistic. Like Emerson, he would live in a commercially booming area of the nation where the new industrial forces were writ large on the landscape.

He'd probably look like Kevin Kelly. Author of the highly oracular books *New Rules of the New Economy* and *Out of Control: The New Biology of Machines, Social Systems and the Economic World,* Kevin Kelly lives with his wife and kids in a bucolic home next to a California state park, where he swaps e-mail, writes, muses, thinks deep thoughts, and keeps bees. He is a prominent boffin among a crowd of "virtual intellectuals," culturati who have moved into digital networks to do their most important work there. Even in this dodgy company, however, he is a standout, because he is so remarkably like Ralph Waldo Emerson. He possesses a solemn, coherent cultural thesis.

Kelly's thesis is that digital networks are opportunity-generating and novelty-generating machines that will permeate the entire constructed world, creating a permanent disequilibrium that is best understood as organic behavior in a technological matrix.

That is an important statement. When the sluggard intellect of the twenty-first century looks from under its iron lids, that is a paradigm it will recognize as its own. If you need a mental box in which to stuff the nature of the twenty-first century, that one fits it best.

So let's carefully unpack that statement, bit by bit. "Digital networks." We have many networks, from railroads and postal systems to gas, water, and electricity, but they'll all become digital, because that's the cheapest way to run them. "Opportunity-generating"—a digital network can pry into almost any situation, opening up new means of analysis and new ways of creating order, which build on one another and can pile up indefinitely. "Novelty-generating"— news spreads fast on networks, and they thrive on pastiche, scrambling old categories and creating mutants and hybrids. "Permeating the entire constructed world"—computation infiltrates everything, as common and taken for granted as electricity and running water. A "permanent disequilibrium"—there is no ideal state for this process. There's never a golden moment in which we can sigh in satisfaction and announce that "the world has been computerized" or "the world has been geneticized." The process of technosocial change just keeps recomplicating itself. It can never be "solved" or "perfected." It has no final aim and no victory condition.

So futurity promises us "organic behavior in a technological matrix"—not a rigid world-machine where form follows function and everything makes good sense, but a world where form can't even *find* function, much less follow it. It's a digital jungle, a world whose institutions are full of blooms and bugs, storms and infestations, algaelike fecundity, sudden death and quick rot. It's a world low in culture, custom, and continuity but high in imagination, ingenuity, and identity crisis.

Kelly's is a rollicking, all-encompassing, Emersonian thesis. It has an unmistakable goofy reek of Transcendentalism about it. It may—it *must*—be very wrong in many important ways, but so was Ralph Waldo Emerson. For better or worse, this is the best summation in print of what's going to happen to us in the twenty-first century. Barring a dark age, that is how tomorrow will look and feel. That is how it *wants* to look and feel. That is the logical extension of the twenty-first century's native driving forces.

Networks with billions of connections and millions of nodes cannot behave like railroads. They roll on where railroads cannot, into intimate aspects of culture, such as writing, painting, and music. Cruising for dates. Religious evangelism. Job hunting. Swap meets. Flea markets. Family gossip. They behave "neobiologically": like swarms, viruses, crabgrass, and buffalo stampedes. If today's children were being taught about how to live in the future, they'd be taught how to live with *that*.

Nobody is likely to teach the works of Kevin Kelly in contemporary elementary schools. Someday, however, my daughter will be middle-aged, like I am now. Assuming she's spared ecocatastrophe and/or weapons of mass destruction, her daily life in the 2030s will look very much more like the giddy prophecies of Kelly's aptly titled *Out of Control* than any school textbook she is assigned today.

Kevin's Olympian assessment is deeply strange, but that is the American way. Most Americans, being Americans, won't even notice that Kevin Kelly is right, even when we're really and truly living in a Kevin Kellyan way, casually sipping our hormone milk shakes and tucking distributed computer chips into our shoes and our doorknobs. To imagine that we'll fret about his big philosophical picture is like thinking that people in Emerson's time spent every spare hour sweating over manifest destiny.

Nevertheless, "neobiological civilization" is the grand trend. Barring catastrophe—most likely, even *after* our catastrophes—that is what will be going on tomorrow. It's going to emerge day by day, marinating us in a hot, rising tide of billions of networked chips.

This situation is the direct opposite of all things minimal, industrial, efficient, and clean. It is tangled, wiry, gooey, and messy. It's a world where machines die much faster than we do, where the recycling smelter is a bigger deal than the Hoover Dam. Most of all, it's not *thought through*. It's all patched together. It's not magisterial and majestic. It's rather squalid and catch-as-catch-can. It looks like what we are willing to pay for, not like what we think might be best for us.

And it's not smart. Computers are getting more sophisticated and much more entangled in our daily life, but they are not getting any "smarter." They're not the ones who are getting a new education: we are.

It was a twentieth-century mistake to think that machines would outsmart us. "Artificial intelligence" has failed. Surrounded by machines moving data, we humans are forced to become permanent students, but the machines cannot get educated. They can be made more elaborate, but they never learn. They don't become intelligent, because they lack the biological, neuron-centered pathway in which a human child masters herself and the physical world.

Unless you have a human body, you don't and you can't get smart. Managing flesh and sensory input—"sensorimotor experience"—is the human brain's main activity. All our other human mental skills have evolved from that rich substrate. Machines lack those skills, and without them their "intelligence" is radically impoverished.

There are still some philosophers and some hard, AI zealots around who deny that thought is inherently embodied. They think that intelligence is based on logical rules. Machines will therefore be made intelligent by first being bestowed with the scholarship, then importing some worldly experience. That approach is entirely against the grain; it is not how intelligence develops. People who try that method are making little progress, so they can't set tomorrow's agenda.

Neurologists and cognition theorists, on the other hand, work with brain matter. They are very busily expanding their frontiers, lighting up human and animal neural action with NMR scanners and radioactive sugar. Real progress is being made in subjects like neural plasticity, long-term potentiation, synaptic transmission. These are the methods by which brains work, as opposed to 1960s clichés of software and circuits. After a century where abstract symbolic analysis held court, the wetware-goo people are coming on strong. They are the ones making the invisible visible.

If you watch children, tomorrow's citizens, with care and an open mind, you will see them educating themselves about physical reality, long before we adults get around to imposing some civilization on them. An infant is extremely naïve, but she doesn't lack brainpower. If adults had brains like infants do, their heads would be the size of peach baskets. Infants have enormous brainy heads, heads too heavy for them to lift.

A human baby learns the flesh—from the brain outward. Sucking is the first order of business, followed by getting the face under some rough control. Almost every conceivable human expression will flit over a newborn baby's face: sneers, winces, winsome grins, baffled fury, world-weary weltschmerz. Long before they master the throat and babble in speech, they babble facial expressions. Infants are energetic and committed learners. They work very hard.

The pursuit of artificial intelligence has been very halting, some say illusory—but it's led to some important revelations. Supposedly advanced applications of human thought,

such as chess and mathematical proofs, are relatively easy
for machines. Seeing and walking, the tasks of infants, are
brutally hard.

We're not conscious of walking and seeing. Until we
tried to get machines up to speed in these areas, we tended to
dismiss these as minor accomplishments, because we didn't
have to consciously think about them. But consciousness is
not the major effort of the human brain. People are not
abstract logical intelligences walking around in big sacks of
meat. People are large animals with big, busy, meaty brains
that sometimes do a few things that are logical.

Once you are armed with this awareness, it changes your
paradigm of learning. When you watch a baby learn to walk,
you can witness the human nervous system mastering the
human body, through a methodical series of titanic accomplishments. It starts with flopping: rolling back and forth
from belly to back, on the body's longitudinal axis. This
painstaking effort is succeeded by an odd squinching
process, where the infant drags its arms and knees below the
torso as weight-bearing members. This achievement is followed by rocking in place. The infant heaves back and forth
on the palms and knees, and sometimes on the nose, toes,
and elbows. It's easy to overlook this vital rocking period,
because it doesn't look impressive—infants don't *move* much
when they're doing this. What the tiny student is learning
here, though, is balance, inertia, friction, and torque.

Then comes the great achievement of crawling. Infants crawl with an impressive variety of speeds, gaits, and techniques. Next, standing. Standing is surpassed by an extensive period of "bridging," in which the upright infant pulls himself along the edges of chairs, couches, table legs, the family dog, whatever's handy. Not only do they learn to stay upright while moving, but they learn to navigate in a complex, harshly punishing world of inert solids. Then, finally, kids do walk.

Walking and seeing—movement and perception—are truly great feats. We share them to some extent with other animals, but machines can't yet do them. If machines ever do somehow "learn" how to "walk," they probably won't deduce the process by any logical or rational analysis. They'll probably mimic the simplest forms of motion, copied from simple animals, such as bugs.

This important self-knowledge should allow us to rid ourselves of obsolete notions of the future as chrome-plated, sleek, efficient, robotlike, and antiseptic. Those are not the inherent qualities of advanced technology: those are the qualities of 1930s modernist machine design. Humans are skin-plated, lumpy, richly intuitive, and very septic indeed. Our biological sepsis is where it's at. Not only are we humans not becoming machinelike; the aspects of ourselves that are least machinelike are the ones with the greatest technological promise.

Technology is not successfully replacing human thought with machine thought. There are no robots and androids walking among us, passing for human. Machines themselves have nothing to teach us. Computers are not marching up the chain of evolution toward the prize crown of human consciousness. If we need a metaphor for technological progression, that's entirely the wrong one. Instead, networks are working like a neobiological jungle, with little dumb things radiating out into every niche. A neobiological technology mimics life. Most living things are bacterial, and most creatures on Earth are beetles. Most silicon chips are really small, really dumb, and don't do much of anything that you'd notice. Technology is getting smaller, sneakier, more embedded, more pervasive.

This is a very important development, but it's hard to describe it as "progress." One might properly object to many of its aspects. But many Americans are far too busy marinating in Kevin Kelly's vision to find it objectionable. They don't complain about it any more than my daughter complains about manning her computer, a form of technodrudgery that she and her friends know as websurfing.

Americans don't even slow down to notice what's happening. They feel at ease out of control—they even enjoy it and export it with gusto. While Kevin Kelly was editing *Wired* magazine, a task he took on for seven long years, the Emersonian "Stockholder Nation" stampeded toward his future as fast as he could scribble. Its citizens have probably never read Kevin Kelly, but they instinctively believe his paradigm, far more than they believe in any outdated twentieth-century form of common sense.

Wall Street's financial aristocracy was properly horrified by price-earnings ratios that bore no resemblance to financial reality. They tried to assert their version of common sense for not seven but *seventeen* years. But since financial reality is not in fact reality but merely a social consensus, a stock bubble that lasts for seventeen years is not really a bubble.

In a networked world in a neobiological civilization, all markets are always all bubbles, a permanent seltzer of bubbles. A Kevin Kelly–style "new economy" doesn't mean permanently high valuations. The only thing permanent about it is that "permanent disequilibrium." A crashed, depressed market has no more common sense than a manic one.

It's a spree for all when the market's ballooning up; if it balloons down, they regret their enthusiasm. But really, there's nowhere else to go. A "flight to quality"? "Safe investments"? What quality, what safety? Networked instability has permeated every niche.

Dot-com companies may die in droves, but so will other companies, companies of all sorts and varieties. Vast oil companies, huge car manufacturers, gritty mining outfits—no matter how analog and tough-as-nails they are, no matter how divorced they may seem from the world of data, they are all vulnerable. Their institutional framework, their control of their circumstances, the discipline of the twentieth-century classroom have been permeated by networks. The networks themselves are as permanent as Emerson's railroads. There is nothing to be gained by rejecting and resisting them. In an American civil war, the side with the most railroads wins.

The new economy that Kevin Kelly describes in his book *New Rules for the New Economy* is not any nicer or more sensible than the old economy. It's certainly not more "democratic," because even if more people trade stock now, conventional voting has less and less to do with what goes on in America. A neobiological new economy might be best understood as Wall Street bulls and bears being replaced by global ants and hornets. This is very novel and extremely troubling. But it is also very American. It may be expressed in daffy technojargon, but there are extensive elements of cultural continuity there. In many ways, it's an Emersonian apotheosis.

So much for the place where our children must find their prosperity. What about our cultural values? Here again we can still learn from Emerson, who promulgated two sets of virtues worthy of the truly educated. These are a fearless adherence to divine truth, and a sturdy democratic scorn for the weary truisms of decadent aristocracy.

For my daughter, these two formulations make very little sense. They would have to be explained to her at length and in detail before she could even work up enough energy to disapprove.

Like any parent, I hope she'll thrive in the years to come. So I would suggest two new virtues for a twenty-first-century scholar: flexibility and patience.

Flexibility because the twentieth century's truisms are in flux or already gone. Flexibility because "persistent" disequilibrium is proving to be, for all intents and purposes, permanent disequilibrium. Permanent disequilibrium does not have to mean unbearable terror, anxiety, and danger. Walking is also permanent disequilibrium. It took my daughter a long time to learn walking, but there was nothing magic about it. By now she's quite expert.

Patience, because she can outlast this condition. The bodies of our children, their first true schools and the true source of the greatest feats of learning, can literally and physically outlive any cultural, commercial, or political change that they are likely to see. If "neobiological" machines "want" to act like ants and gerbils, that means machines will die like ants and gerbils, swiftly and mostly invisibly, under the bark and the grass. But the large persistent flesh of today's children will not die for quite a long time, even if their attitudes and skills date quickly. Today's children can physically outlive almost any vogue for any "new" economy, unless they somehow happen to be physically killed.

Therefore, be patient, and governments, institutes, corporations, entire industries will be dragged dead past your tent. You need not quail in terror that the very bedrock of society will shatter from technical change. Because there is no bedrock. You don't need one. When you leap headlong from the flaming windows of your parents' former certainties, there really isn't that much difference in your circumstances.

There is no gold standard for civilization. You don't have roots; you have aerials. Henceforth, O children, you are going to live in a world glued together by networks. Networks consist of two things: connections and nodes. Connections are temporary and flexible, while nodes are persistent and solid. You are the node; your circumstances are the web. You should treat the connections with great flexibility and the node with utmost care and respect. Flexibility and patience are the two virtues that best suit those circumstances. If you are handled with care, you will outlast everything that you understand. Your own children will not understand your reminiscences, but you might as well accept that philosophically; you'll be getting as good as you gave.

Let me end this chapter, the student's chapter, as a teacher ends the school year: with a high-minded exhortation. Children, school is boring—I know that. But you shouldn't mistake the stifling routines of scholastic discipline for the process of learning. If you can make enough appropriate noises to pass out of the school and avoid being placed in jail, you'll be doing plenty of learning once you're outside. You're likely to thrive if you learn plenty about subjects where the tests and grading papers have yet to be invented. And if you find yourself learning about something unusual and there's no sense of drudgery to it—on the contrary, you find yourself spending long, smiling hours just painlessly soaking it up—take my advice and look for a job there. If there don't seem to be any jobs there, find a way to make one up. Plenty of your contemporaries will also be doing it. Be sure to take the time to learn from them, as well as the usual geezers.

Twenty-first-century people will spend a lifetime learning, and it's a blessing and a curse. It might well be argued that there is something inherently unworthy and servile about adult people having to learn all the time. There you are, in your future, half slumped over your keyboard, living among jittery, faddish networks that are out of anybody's genuine control. Learning all the time is kind of a drag. It's rather downbeat, humbling, and restrictive. You don't get to be a Homeric hero, or a Muslim saint, or a Nietzschean superman. Flexibility and patience aren't heroic qualities. They are good qualities for mice in the woodwork, creeping little mammals in a jungle of giant machines.

But the machines are not the giants in your future, kids. The machines are the ants. The machines are much more temporary than you are. They're not getting bigger, they're getting buggier. If you want to be the giants, that's up to you.

STAGE 3 THE LOVER

And then the lover,
Sighing like furnace, with a woeful ballad
Made to his mistress' eyebrow.

My heart skips a beat when I contemplate this chapter. All the world loves a lover. We can't help but be touched by his hot-breathing ardor. Plus, his eyebrow obsession provides comic relief.

Romantic love is the best example of "philosophy in the flesh," the needs of embodied intelligence defining our place in the universe. When we're in love, the distant moon and stars smile down on us. Every random song on the radio addresses our concerns. Piano legs look sexy. We'll even plead with that eyebrow as if it were the body part making the executive decisions.

In the nineteenth century the critic John Ruskin invented

a very useful concept: the "pathetic fallacy." We humans commit the pathetic fallacy when we project our human feelings onto symbolic externalities. The pathetic fallacy is a confusion between the powerful way we feel inside and the indifferent way that the world actually works. Ruskin's most famous example is a dead girl in a rowboat. "They rowed her in across the rolling foam—the cruel, crawling foam." Grief has so stricken the bereaved lover that he feels that even sea foam has an agenda: the crawling foam is cruel.

The pathetic fallacy is a fallacy because, although lovers are responsive, the physical world is inert. The material world is cold clay, like the dead girl in the boat—it cannot respond to our needs and feelings.

Or at least it didn't used to respond. John Ruskin was the greatest design critic of the nineteenth century, but this is not John Ruskin's time. A rather more contemporary figure, eminently suitable for the twenty-first century, is Professor Neil Gershenfeld of the MIT Media Lab. In 1999 Professor Gershenfeld published a remarkable book entitled *When Things Start to Think.*

In Gershenfeld's futuristic laboratory of cyberneticized physical objects, the dead things around you *can* remold themselves to suit human needs and feelings. Suddenly, it's as if all the world loves a lover. Shoes, hats, bricks, milk bottles, the fridge, the sink, the phone—everything's got a chip inside, and everything's got an agenda. When things start to think, then the pathetic fallacy is no longer fallacious.

With "ubiquitous computing," the world becomes your darling! Bits seduce and juicily mingle with atoms! The cold clay twitches, opens camera eyes and microphone ears! What was once a distant, glassy "interface" becomes a hands-on, fleshy, sexy "interaction."

So across their gap of centuries, Gershenfeld and Ruskin have a grand common theme: a fierce engagement with materiality. Ubiquitous computing is Ruskin's pathetic fallacy in a high-tech world. "Ubicomp" (as it's coming to be known) is a new and scarily intimate relationship between humanity and its material surroundings.

Mind you, "things that think" do not love us. They do not even think, strictly speaking, much less emote or write sonnets. Lacking embodiment, they have none of the sensorimotor skills that do most of the brain's real work; lacking hormones, they have no desires. So a "smart" desk stapler cannot return my feelings.

However, a chip-jazzed device is certainly in a cruelly fine position to do all the awful things that drive us crazy in dysfunctional love affairs. It can, for instance, drop dead when I really need it and depend on it. It can whine crazily for my attention, intrude on my privacy, even stalk me. Unicomp is a twenty-first-century technology that poses intimate, touchy-feely opportunities and problems. Suddenly, almost every object around us is capable of failed commitment, heartless exploitation, and sleazy personal betrayal. Can't live with it, can't live without it.

John Ruskin, to his sorrow, could never live with love. His wife left him because he couldn't bring himself to overcome his hypercritical squeamishness and consummate their marriage. So he fell to his own fallacy: he spurned a living human being and preached "faithfulness to materials" instead.

Ruskin failed as a husband and became the supreme guru of the Arts and Crafts movement. This visionary act of Ruskin's has some profound lessons for the twenty-first century.

Ruskin had extremely firm ideas about materiality and how it should behave. The Arts and Crafts approach was a passionately engaged, demanding, intimate relationship between humans and material goods. According to Ruskin, objects and buildings should reflect their local culture and their local character, instead of getting above themselves in a jumped-up life of coquetry and crass pretense. Structures and objects should be built to last and should age gracefully. They should never, ever have flimsy, fraudulent veneers or the smeared, slutty makeup of stucco. Household goods should never cheat people by concealing faults in their construction. The objects cherished in the Arts and Crafts household should be frank and forthright: useful, beautiful, or both. No knickknacks, gadgets, lingerie, or fabulous funky fetishes.

Like romance and marriage, Arts and Crafts has failed

repeatedly. Its spirit is very persistent, however. The twenty-first century is quietly dissolving many of the roadblocks that always spoiled that romance.

Since the dawn of mass production in Ruskin's day, the finer feelings of people of taste have consistently been betrayed by cold, crass industrial clumsiness. Ruskin's language of protest may be old-fashioned, but his indignation is thoroughly up-to-date. Every generation rediscovers the harsh truth that our material goods are grossly unworthy of our feelings. So every time Arts and Crafts is dismissed as corny, impractical, softhearted, and mushy, it reappears in some vigorous new guise.

Anywhere that shopping looks like art and philosophy, there the lonely ghost of John Ruskin moves among us, giving his aching heart to the rough-hewn beams, the weathered slates—the lumpy wooden kid toys, the Guatemalan jackets, the bamboo wind chimes—to all things organic, warm, rustic, honest, sincere, and above all, to things that are committed and faithful.

The dawn of the twenty-first century is a golden age of design. It is a period of disposable cash, flourishing designers, radically innovative goods, and an ardent and worshipful consumer base of ultrainformed connoisseurs.

Our historical situation is radically distinct from Ruskin's. There are technical feats we can accomplish that were beyond his imagination. Our materials aren't natural,

local, and authentic, like Ruskin's were. They are fantastically ductile, global, and increasingly infested with pro-
grammed interactivity.

We now manufacture a great many of our worldly goods
out of "cruel, crawling foam." Crawling foam, more formally
known as "injection-molded plastic," is a substance of tremendous significance, stunningly well suited to computerized design and manipulation. Injection-molded plastic is
the least honest material that the human race has yet
invented.

It is impossible to be faithful to injection-molded plastic.
John Ruskin was so picky that he couldn't abide even cast
iron—he considered it too gooey, liquid, and cheap, suitable
only for madhouses. But today we have, along with all those
old-fashioned "modern plastics" of the 1960s, a truly weird
repertoire of ductile goos such as Styrofoam, nylon, Kevlar,
urethane, epoxy, and silicone. These materials are so profoundly artificial that they are simply beyond the ken of
Ruskin's ideology and cannot ever be recuperated.

The crawling foam has very few design limits and almost
no material authenticity. It is not local to anything or anywhere, and it can have any property that you are willing to
pay for. Everything about it is arbitrary: its grain, texture,
color, weight, shape, elastic modulus, strength in compres-
sion—all these qualities can be specified on demand.

These new powers and new materials have led to the
design and creation of a singular totem object of contempo-

rary times: the "blobject." The term was coined by contemporary designer Karim Rashid, author of a book aptly titled
I Want to Change the World. "Blobject" may sound like a
comical bit of design jargon, but with a little coaching, one
can learn to see that blobjects are thriving in fantastic numbers and littering the modern landscape. They are computer-
modeled objects manufactured out of blown goo. They are
rounded, humpy, bumpy plastic creations. They are often
translucent. And though they're merely made things, blobjects tend to be fleshy, pseudo-alive, and seductive: rubbery,
grippy, flexy, squeezy, pettable, and cuddly.

Some contemporary examples: the Gillette Mach 3 razor.
The Oral-B toothbrush. The Swatch Twinphone and the
Phillips USB desktop video camera. The Handspring Visor
PDA. Gelatinous wrist rests. TechnoGel in office seating.
"Morph" Cross pens with bulbous gel grips. The curvy, slith-
ery Microsoft Explorer mouse. The curvy, plastic Oh chair
and magnesium Go chair.

Automobiles were among the first objects to enjoy the
benefits of computer modeling. Taillights, windshields,
hoods, and fenders often look strangely rounded and ductile
nowadays, as if they had grown in place or had partially
melted. The New Beetle, with its humpy retromodern look,
looks like it was cast in aspic.

Blobjects carry the flag in a world whose manifest destiny is "organic behavior in a technological matrix." Chips
shape them and make them behave. Computer-aided design

and injection molding allow them to assume any form. They get their organic forms directly from us: from mimicking human flesh.

Unlike classic twentieth-century industrial objects, their form does not follow their function. That's because their functional parts, being chip-based, are too small to see. Form can no longer even see function, much less follow it. Since blobjects are made from molten goo, they can take on any shape, cheaply and dependably. So they have adapted themselves to the only remaining design limits: the sensori-motor needs and desires of the human body.

Blobjects evolve from human ergonomics. Screens must be large enough for human eyes to see. Buttons must be properly sized for human fingertips. Telephones must be properly sized to the human mouth and ear. The ergonomi-cally spectacular Oral-B toothbrush is all about human teeth and human gripping pads, right down to the crook of the little finger.

Blobjects look like us because they stick around with us, and they live with us, and they try to please us—they are blobs because we're blobs, too. We also grow, we're spongy, we're curvy, we're pettable, we're eager to please, and some-times we're even lovable, if people can only see us for who we are, swallow their distrust and distaste, and give us a chance to make them happy.

Blobjects are not impressive industrial Molochs like the steam locomotive or the Saturn-V rocket. They are humble, disposable, and easy to miss. They resemble commensal organisms, little remoras or pilot fish, snuggling up to the human body, attaching themselves to belts and sneaking into purses and backpacks. They puff up like rubbery puffer fish to seal human feet with air pressure. They even worship our eyebrows as tinted designer sunglasses.

Translucency has become their significr of digital power. Translucency says that they do something strange inside themselves, something potent but hard to see.

Blobjects are intimate and disposable: they don't deal in permanence or monumentality. There are no great architectural blobjects, though the rippling metal extravaganzas of Frank Gehry come close. Future Systems of London has an unbuilt skyscraper project called the Blob, while Lord Foster of Thames Bank has riposted with a master plan for the Gherkin. Although blobjects creep and swarm about us in their very millions, they don't dominate our skylines, and for good reason.

This age is quite prosperous, and great things are being done, yet just as we have no proper way to conduct a love affair and no proper way to be married, we have no proper way for our buildings to look. Our period has no signature design style. The structures we build are reheated Arts and Crafts, reheated neoclassical, reheated Bauhaus, reheated everything, really—all the looks of the past, cut up and quoted all at once.

Blobjects don't offer us much help here. Very few people

would choose to live in a fully blobjective environment, a wobbling inflatable tent furnished with beanbags. Contemporary people strongly prefer to live among ritual architectural symbols of personal heritage and continuity. These pastiche forms ("colonial," "classical," "Southwestern," "Tudor") used to be stark necessities from the past's construction industries. Houses with these shapes are no longer technically necessary, but they hang around because of legislation, property values, and sentiment. These forms are technically known as "skeuomorphs," old shapes patiently carried into a new culture because they bolster our self-image, define our identities, and somehow make us feel better about ourselves. Skeuomorphs are tastefully retro and carry a reassuring message of firm purpose and continuity.

Blobjects don't offer us these ritual comforts, because blobjects are the unreal thing. They are the genuine avatars of contemporary technological circumstances, which are formless, full of opportunity, ultraflexible, and radically flimsy. Blobjects have something genuinely unsettling and uncanny about them. Their lines are sometimes Pokémon-cute, but they're commonly fungal, epicene, and creepy.

Although they look vaguely "organic," they are profoundly unnatural, because they come from computer-aided design and manufacturing. CAD-CAM is a technological matrix of mathematically specified computer geometries. Blobjects are really new, a genuinely modern design trend, in the way that the Art Nouveau whiplash line was once new, in the way that streamlining was once new. Because they are small, temporary, and throwaway, they incarnate the spirit of the times.

This computer in front of me (a poison-green Apple iMac) is a blobject. It's also the center of my literary life; it's a blobject that I'm quoting John Ruskin on, that I was using earlier to pick a minor fight with Robert Louis Stevenson, that has structured a book around a Shakespeare soliloquy. The machine assembling this text is translucent, "fun," "designery," "user-friendly," "ergonomic," and rather pricey, and it also has a tragically short life span.

Nevertheless, I have a rather serious relationship with this device. I spend many hours in its company, and I earn my living with it. Of course it cannot sincerely and faithfully return my feelings. But it does have its own name. It's demanding and temperamental and very unforgiving of abuse. I'm forced to pamper and pet it rather more than I do my cat.

More to the point, since it's networked, it is a relationship machine. My beloved wife and I boast a relationship that has gloriously outlasted whole generations of computers. She is by no means a corny "computer widow." We *both* websurf constantly, and we send each other supportive, helpful, spousely e-mail through the Internet—even though we and our dual computers are in the same house.

Art Nouveau once produced a profusion of sleek, curvilinear, handmade art objects. No mere thing had ever looked that weird and ripply before, but Art Nouveau was a counterculture, powered by strong feelings of rejection and reaction against a mass-produced industrial society. Its zealots chose to make their objects look that way in order to show up the crass, square, clunky products of early industrial factories.

Art Nouveau's ripply art vases and drippy mirror frames may seem best suited to people in egret feathers and tiaras. But blobjects are Art Nouveau products' legitimate heirs. In fact, blobjects are their avengers. Blobjects are almost as ripply, whippy, and hysterical as Art Nouveau handicrafts at their most extreme. However, blobjects are not rebels against industrialism—they are the postindustrial conquerors. They survived, killed, and buried twentieth-century mass production. They are fully tied in to the dominant industries of the new century: computers and networks.

The well-to-do bourgeois intelligentsia of an information age have a lot in common with the wealthy patrons of Arts and Crafts. They don't dress for dinner as Edwardians did; they don't keep footmen and use a full set of family silverware. But they are just as extravagant in their own way; they will lavish any amount of money, care, and attention on "tools." I can see a swarm of tools around me as I type this. My iMac is a tool; my digital wristwatch is a tool; my Teva sandals are tools; my Kensington turbo-ball is a tool; so is my boom-box stereo and my Artemide e-light and my ergonomic office chair. Blobjects love to pass as tools. But they really thrive best when they are tools and entertainment at the same time.

Edwardian status signifiers were horse-drawn carriages and high-button shoes. Ours are direct period equivalents, such as the sport utility vehicle, which is a fun, luxurious military-spec truck. And the cross-training shoe, which is high-performance athletic gear for daily office wear. Our possessions often seem best suited for the harsh demands of a mountain-climbing rescue squad, even though we use them to go get a cappuccino.

Modern devices are overstuffed with functionality for two good reasons. First, because functionality is cheap and easy in an age of chips. Second and more important, because function-as-baroque is making symbolic promises to us and to other people, ardent promises that we very much want to hear from our toys, tools, and appliances.

The goods produced by Arts and Crafts made an identity statement for the owner: I am sincere and committed; this is where I take my stand to defend my values. But a device with huge amounts of extra functionality says something very different: I will not be trapped and cornered by technical limits.

Arts and Crafts objects and twenty-first-century blobjects have a great commonality. They are both elitist, and they both despise consumers. In their eyes, consumers are faceless, tasteless creatures of mass industrialism, mere passive drones with false consciousness hammered into them by commerce and bad media. Consumers buy whatever cheap good is most loudly promoted by hucksters.

But blobjects are not made for the consumer. They are made for a knowing participant in the technocratic scramble for wealth, the "end user." The end user does not consider himself a consumer. He's not plucking random junk off the wire rack for the sake of having more stuff. He is deeply engaged in the system, exploiting and adapting it. End users are a brainy, fastidious, postindustrial ruling class.

There was once a gray-flannel "consumer society," but its day is gone. Half a century has passed, consumer society has been transformed, and the process is ongoing.

But this kind of sociology always sounds remote and abstract, like referring to a kiss as "courting behavior." Let's try to get completely hands-on here. Let's cozy up to the warm and sensitive place where human flesh physically touches the network.

We can make this distinction clear by bringing some Ruskinian critical clarity to bear on certain telling details of design. Let me show you how two quite similar devices reflect two very different societies. We'll take some time to critically examine two telephones: the Henry Dreyfuss Bell telephone of 1950 and the Motorola StarTAC cell phone of 1999.

Henry Dreyfuss (1904–1972) was a titan of America's consumer century. His Bell desk-set telephone, the Model 500, is a design icon; industrial designers still speak of it with hushed respect. A masterpiece of form and function, it came in Henry Ford basic black. It sat forthrightly on its square, heavy, sculpted base, and it had a dial, a grip-friendly handset, and a curly cord. This telephone was simple, direct, integral, and honest. The Model 500 is still instantly recognizable as "a telephone"—in some sense, it is *the* telephone, and its popular image can never be superseded. More than ninety million of them were manufactured.

Dreyfuss's telephone is not merely mechanically efficient. It has been deeply thought out, and designed for a specific social role. It gives the consumer a firmly assigned place in the industrial order.

None of the functional parts are visible. For customer-friendly convenience and monopoly dominance, they have all been "cleanlined" away, beneath the shell. So there is no need for any user documentation, user's manuals, or any specialized training in the use of communications networks. The phones work very well, and they all work exactly alike. They're so legendarily sturdy that they work after falling out of four-story windows; they even work after the house catches fire. Marginal social figures such as little kids, ailing grandparents, and foreigners have no trouble using them.

Now let's consider the contemporary Motorola StarTAC cellular phone. I happen to have one right here in my jeans pocket, since it is a blobject and it commonly clings to my rump. I'll hold it in my hand and we can all have a good, close look at it. In fact, if I had your number, I could talk to you right now. Instead of my words boiling off from ink and paper with a lag time of months or years ("Yesterday Now"), you'd have a lag of mere milliseconds; you'd have me clamped to the side of your head; you'd hear me breathing into this thing.

This device is created for someone who clearly *resents* an assigned place in the industrial order. Anyone in such a stodgy, static position nowadays runs a grave risk of being swept aside. End users are the evolved form of twentieth-century consumers. They have been trained by years of economic upheaval to search for open-ended skill sets that can ensure their employability. They feel most comfortable when they have plenty of maneuvering room in which to exercise technical mastery and maintain their elite status.

So while a Dreyfuss phone merely says 2-ABC, 3-DEF, and so on, a StarTAC blurts a whole technospeak litany: BATT-LOCK-MUTE-RCL-STO-CLR-SND-PWR-VOL-FCN-END. Though it's small and flimsy, this phone comes with three-digit lock codes, six-digit security codes, multiple service-level restrictions, multiple display languages, nine different ringer styles, and even a scratch pad.

Although I own a Motorola StarTAC (for this passing moment, that is; by the time this book sees print, I will likely have dumped it, along with the iMac computer that wrote this text), I'm what is known to the cell-phone trade as a mere "safety user." Unless I'm out of town, I mostly use it to coordinate family matters: to recite the menus at take-out places, for instance. So, according to twentieth-century consumer-design philosophy, I should be much happier with some much simpler, more user-friendly device. A phone that carries out simple, specific functions in a predictable, rugged, single-minded way. A device that, like the Dreyfuss phone, "solves the problem."

But end users don't want to solve problems. A solved problem is actively dangerous for them. Any end user with a permanent solution has lost a job. End users are postindustrial people. They survive by creating new opportunities and by networking feverishly inside unstable, rapidly changing industries that are in permanent disequilibrium.

That also explains why end users don't settle for cheap, simple, fully usable software. After all, if software is simple and usable, then anybody can use it. End users are in meritocratic competition over technical jobs. They can't afford to be just anybody, because this is a swift ticket to poverty.

The grand trend in the information revolution is for all the simple, predictable, solvable activities to move toward the poorly educated and poorly rewarded—or on to machines. So skilled, high-paying jobs become progressively more and more like the apex of human difficulty—in other words, management politics. Instead of bending metal every day in the machine shop, you spend your time negotiating on relationship machines, such as this cell phone. Coordinating. Compromising. Building coalitions. Answering your phone mail and reading your e-mail. Getting and staying up to speed. As a general principle, the more office e-mail you're reading, the harder you are to replace.

But toys, tools, and devices aren't built just for the sake of their buyers. We must also bear in mind the pressing needs of the manufacturer, whose interests often directly conflict with the customer. The cell-phone companies do not want to "provide a service." Once a service becomes a predictable commodity, the "problem has been solved." This means that there is a swift competitive race to the bottom, and profit margins become razor-thin.

Therefore, the cell-phone company sells me a StarTAC at a loss. It does this in order to entice me into a committed, networked digital relationship. The sexy hardware is merely a come-on; the relationship is what matters.

The cell-phone company certainly doesn't want me pursuing any simpler, purer, easier way of life. Not only will I go broke if I do this; so will they. So they want me fully *engaged in the process;* they want to keep my eyeballs faithfully stuck to the screen and my fingertips caressing those little phone buttons for as long as possible. After all, they are selling me hours of use—not efficiency. People with simple, efficient, solitary lives don't much need cell phones.

So the service provider wants end users to network and to keep paying money and attention. End users want an innovative environment in which they can move up and compete, without being sidelined and bankrupted by obsolescence. The powerful vector between these forces, these fierce demands from the end user and the manufacturer, has given birth to this new kind of device. It is a ductile, very capable, extremely demanding, disposable relationship machine.

That is why this Motorola StarTAC looks and behaves the way it does. Now I can put it back in my pocket, where it can once again disappear and be taken utterly for granted.

These forces are not the only forces possible in industrial design. Arts and Crafts objects of the Belle Époque quickly vanished with World War I. Military and security issues trump capitalist and consumer ones, at least during the short term of actual shooting war. A paramilitary cell phone for "safety users" in unsafe times would look very different. It would probably still be a blobject, but it would be a GI blobject, taking on the tough, government-specified performance aspects of an aircraft black box. When wartime faded, however, a GI cell phone, too, would look corny and would vanish from the reviving marketplace, chased off like army boots or a civil-defense helmet.

As lovers know (or quickly learn), an ongoing relationship is not a "problem" that can be "solved." A committed loving relationship is multiplex and time-bound, involving long biological cycles of growth and decay. Married people have a thriving host of problems, but if they "solve" them, the family dies.

Postindustrial blobjects don't "solve problems" either. Many of the signature devices of our times are portals into long, complicated relationships with service industries. They are cell phones that sell hours, boom boxes that sell tapes, laptops that sell software.

These are not anonymous consumer transactions, in which I pay some cash, take my box off the rack, satisfy my dark lust for possessions, and am henceforth left alone. All these devices demand continued, meticulous pampering and attention. They are not "consumer" products. They provoke and demand end-user behavior. This is close in spirit to the hands-on, intimate, ticklish Arts and Crafts relationship of an artist and a patron. Except that this patron-client relationship is mediated by machines. Very demanding, very observant, fickle, temporary, faddish machines.

Machines of this advanced variety really need a new name. The old industrial term "machine" has far too much cultural baggage; it has the antiquated semantics of Charlie Chaplin's *Modern Times*, where the Little Tramp is folded, spindled, and mutilated by huge, blind assembly devices. That Chaplin situation was not a permanent dystopia; it was only a temporary stopgap.

In our post–*Modern Times*, it makes no commercial sense to use expensive human flesh to carry out dumb, simple, repetitive actions. The insufficiency of machines was once a human job magnet of titanic proportions, but that era of merely mechanical insufficiency is over.

So contemporary people are not inhumanly degraded into assembly-line robots, as in the sci-fi visions of *Modern*

Times and *Metropolis*. Not because of noble sentiment or social justice but because machines have become far more ductile and disposable than human beings are or ever can be. The movement is all in the other direction. Machines are taking on organic qualities. Not intellectual or spiritual qualities, even though our philosophical tradition considers this the great distinguishing mark of separation between humans and machines. Machines just aren't that good. They can't see, they can't walk, they can't feel. They are apprentices in organic behavior; they can't jump right for the top of the food chain. They are taking on *primitive* organic qualities: buglike and petlike qualities.

So "gizmo" is a better, more truthful term than "machine." Because a gizmo is a small, faddish, buzzy machine with a brief life span. Some gizmos are blobjects. Most blobjects are gizmos. When you buy a cool new blobject gizmo, you are at the hot-wired cutting edge. You are living large!

The nineteenth century made machinery. The twentieth century made products. But the twenty-first century makes gizmos. In a "product," form follows function. There isn't much decoration, because that would be irrational and inefficient; it increases production costs on the assembly line. For a gizmo, the function *is* the decoration. A gizmo, like a cell phone or a jogging shoe, has more functions than the user will ever be able to master, deploy, or exploit. It is designed to have baroque and even ridiculous amounts of functionality. A gizmo "empowers the user" but not in any permanent or predictable way. It has irrational levels of power, which are based on experiential values like "fun" and "amusement" and "involvement" and "technical sweetness" and all things hip and designery.

A gizmo is neither a "machine" nor a "product." It doesn't want you to accomplish any task in particular. It wants a relationship; it wants to be an intimate experience, as close to you as your eyebrow. It wants you engaged, it wants you pushing those buttons, it wants you faithful to the brand name and dependent on the service.

A gizmo needs an interface, and an interface for its interface. It needs tech support, and tech support for its tech support. Even its web pages need web pages. And this is where you work. Because the *mental* insufficiency of these bleeping, begging little gizmos has become a human job magnet of titanic proportions. The near-infinite complexity of a network of rapidly obsolescing, disposable gizmos can suck up near-infinite amounts of human effort and ingenuity.

This situation may sound frivolous, unserious, even contemptible. But wait a minute: imagine if life were otherwise. Try to imagine John Ruskin's version of an Arts and Crafts computer. Imagine that you were permanently, faithfully committed to one computer that was handmade of local stone and wood instead of merely carrying out a cheap, sleazy, temporary infatuation with a piece of plastic junk. Just suppose that some ultra-Ruskinian World Design Council "solved the problem" of computers, once and for all.

They invented and mandated the last, perfect, faithful, authentic computer. So that you were given just one, a permanent one, for your whole life.

Think about that; think how terrifying and tyrannical that relationship would become. You would have to guard that machine with your life. If you weren't simply ground underfoot by Orwellian surveillance, you'd probably have to learn machine language and assembly code. Since divorce is impossible, any misstep would be fatal, the key to a lifetime of misery.

But since it's just another throwaway Dell Wintel piece of junk, you still have room to breathe. Mind you, the computer is very insistent about its needs for expansion and upgrades. It still wants its mouse cleaned out and its itching viruses scratched. But it's not entirely stifling, tyrannical, and codependent. So you might as well break out into the full, happy gizmology of screen savers, streaming MP3s, wallpaper, virtual pets, and so forth.

A computer is an ultimate gizmo because it is both amazingly powerful and amazingly temporary. It follows that anything mediated by a computer, anything containing a chip, tends to take on a gizmolike character. A car with a computer is a computer with wheels. A plane with a computer is a flying chip. Pretty much any object or any process can contain a dedicated chip. This should mean that a gizmo carnival is just over the horizon.

The holdup is not the chip or the wireless communications technology. Chips like Transmeta and wireless systems like 802.11 are clearly moving in this direction. The holdup is batteries. We just don't have good portable power systems. We can't plug everything into everything, because wiring is both a leash and a trip hazard, and some objects must be portable to function. This lack of portable energy means that the army of chips already installed around us spends most of its time inert. So many objects that might plausibly have chips, such as forks, hinges, and bricks, are left undisturbed in their medieval Ruskinian peace.

However. If you had a very small fuel cell that would run quietly and dependably, a kind of warm, primitive gizmo heartbeat, that would change things radically. Your house would swiftly be infested with lovable commercial Pokémon and Furbies. An overwhelming plague of Furbyization would hit everything you own: toasters, vacuum cleaners, jogging shoes, your television, your pet's collar, beer bottles, aftershave, deodorant, toothbrushes. There would be a silent cacophony of interactivity in all domestic and industrial objects. Should this happen, the source of Ruskin's pathetic fallacy will finally rise from her rowboat, and our world will become uncanny.

Ubiquitous computing is still speculative, but it's not imaginary. It's receiving a great deal of serious study in serious places such as DARPA, IBM, MIT, Matsushita, and Motorola. The sponsors of research into tiny fuel cells are also a host of industrial giants: 3M, Dow Chemical, Eveready,

Lucent, Dell, Gillette, even Procter & Gamble—a whole panoply of industries, every one of them dead tired of lugging dead leaden batteries around, every one of them throbbingly anxious for gizmos that can wake up at a kiss.

There are many names for this ubicomp concept. The jargon of the infant industry is young and still unsettled. Wearable computers. Intelligent environment. Wireless Internet. Peripheral computing. Embedded Internet. Ubiquitous computing. Things That Think. Locator Tags. JINI. Wearware. Wi-Fi. Personal area networking. And so forth. This kind of disruption in the English language is like the rumblings of a tectonic fault. The signs are good that something large, expensive, and important will tear loose there.

Suppose that this really happened. Never mind what name it had—what would it mean and how would it feel? Motorola is a sponsor of the Things That Think effort at MIT Media Lab. It likes to call its version of this concept "digital DNA." This rhetorical notion—DNA inside machines—fits very well within the grand theme of "organic behavior in a technological matrix."

But the first suggested uses for ubicomp are still primitive: chips that are tied to larger devices, providing their power. A refrigerator, for instance, is always plugged into the wall. So perhaps a "smarter" refrigerator could read the bar codes on all the goods that enter and leave it. It would then "know" that you had no milk. Perhaps it could order some milk for you off a website. Or it could answer its cell phone when you called it from the grocery, and get you up to speed on its contents.

Cars also have plenty of onboard power. So a car might as well become a mobile office cubicle; it will talk to a cell phone and laptop, read text files aloud over its radio speakers, take phone calls, even ask for handy directions from satellites overhead. The smart tires will complain when the tread gets low. The smart gas tank knows all its favorite gas stations in the area. The speedometer silently calls Dad whenever the family teenager does ninety miles per hour.

These ideas are clever, but they're far too limited. They just add a sexy blink and smile to products that already exist. They don't yet express the radical change in the intimate relationship between humankind and things.

The chance of heartbreak here is very grave, because if physical objects misbehaved as badly as most computer software does, human life would become hellish and possibly murderous. It is scary and profoundly unsafe to hook physical processes and events together in unpredictable, invisible ways. Some pervasive computing notions clearly have the calamitous potential of an *I Love Lucy* episode. Misreading a bar code, the refrigerator orders twenty gallons of beluga caviar. The smart car follows directions in the rain and darkness, then routes the unwary driver straight off a broken overpass. A single instant's bad driving can kill you. People properly tend to be very conservative about kitchens, which can slice, dice, and fry the unwary.

An area of stronger promise is express shipping. Here we have a ubicomp scenario that is already successfully unfolding in real life. With the modern express package, chip function is added to a portable object in a way that is not only convenient but a definite competitive advantage. I can follow a package via Internet from distant New York right to the doorstep of my business.

If I could keep that schedule for all raw materials that down-to-the-minute, then I could reschedule my inventory, keep stockpiles low and lean, do just-in-time assembly, and make a whole lot of money.

I don't much want to "talk" to a package. I don't want a package tugging my sleeve, stalking me, or selfishly begging for attention and commitment. If a package really wants to please me and earn my respect, it needs to tell me three basic things: What is it? (It's the very thing I ordered, hopefully.) Where is it? (On its way at location X.) And what condition is it in? (Functional, workable, unbroken, good to go.) The shipping company already needs to know these three things for its own convenience. So it might as well tell me, too.

So the object arrives in my possession with the pathetic fallacy attached. A cute meeting, a new relationship, O happy day! The future can break loose in a rush. When that object arrives, I keep the tracking tag.

Let's say that it's a lovely anniversary gift: a handsome Lava lamp. Though I'm touched by this sweet gesture, I really don't need that thing every day. It might be kind of fun during a party, but by then I've already fatally scorned it. Like the moldy spouse of Mr. Rochester, it's settled into dank neglect in a gloomy attic.

But suppose the lamp still has that shipping chip. That means the lamp answers when called. I just look up its location on my home tracking network. The gizmo keeps the faith by responding to my three basic questions: (1) it's a Lava lamp; (2) it's in the southwest corner of the attic; and (3) it still works fine. Out it comes, petted and shined, ready for its moment of glory.

Having benefited once or twice by this sound, mutually fulfilling relationship, I take the logical next step. I tag all the toys, tools, blobjects, and gizmos that I already own. Now I have a menagerie, a veritable Ottoman harem of smart things, but they don't quarrel and bicker. They are faithful and disciplined, speaking when spoken to. They are digitally ranked and serried, fully in touch with their inherent thingness.

Huge benefits ensue. I no longer need to sweat and struggle to put my possessions into order. My things can never get lost or misplaced. They can't even be stolen from me, because the tags are too small to see and will avenge me on the thief. Best of all, when they become garbage (as all gizmos are inherently likely to do), they are *smart* garbage! In a dutiful ecstasy of self-immolation, they identify themselves to the junk recyclers, who swiftly arrive to bear them off to the Almighty Smelter of the Universal Ooze, back to the crawling foam.

They fold themselves right back into the production stream. This organicized production matrix doesn't spew toxins or waste materials, because even the trash will think. Cue those horns and violins, maestro! Romance sweeps majestically into stable, connubial bliss! We all live happily ever after!

Sounds lovely, doesn't it? My goodness! But wait—even though Stage Three: The Lover is ending in a radiant, golden bliss of happy-ever-after, history still hasn't stopped.

The Seven Ages of Man must roll on. We're now a mere page flip away from the far, far more sinister world of Stage Four: The Soldier. What on earth could that mean? Suppose that the things around me weren't made by people who adore me, study my habits closely, beg for my attention, and obviously want my money. Suppose that cell phones were best suited for digging my corpse out of rubble, and that fast shipping systems sent me deadly contagions from other continents. Suppose that my world was a militarized world, full of implacable fury and ruinous bitterness. A world designed not by lovers but by my enemies, the people who hate me, mean me harm, and want me dead.

THE SOLDIER

> *Then a soldier,*
> *Full of strange oaths, and bearded like the pard,*
> *Jealous in honour, sudden and quick in quarrel,*
> *Seeking the bubble reputation*
> *Even in the cannon's mouth.*

Judged from modern Washington, the capital of the world's only remaining military superpower, Shakespeare's soldier is a fanatical lunatic. Feuding with your own comrades to make a reputation for valor. Jumping into cannon fire because it's honorable. What is this behavior supposed to gain anybody, except for prison or Prozac?

The U.S.A. is a great military empire—with a difference. The contemporary American military doesn't conquer real estate. The Americans are focused on sea-lanes, pipelines, airstrips, and oil resources. They build global refueling bases and electronic listening posts. They have extensive military assets in orbit.

Historically, from Ramses to Stalin, the point of warfare was annexing enemy territory. But the American military considers this a loathsome chore. In Shakespeare's day, Sir Francis Drake robbed treasure galleons and put Spanish towns to the torch. But American buccaneers do not jump from their aircraft carriers to loot and burn Kuwait.

Looting is preindustrial behavior, while America is a postindustrial empire. The Americans can derive no long-term profit from looting, because looting harms the means of production, distribution, and information. If the American military sacked and looted anybody, the U.S. State Department would have to return next quarter with the World Bank and the International Monetary Fund and shore everything back up in order to reassure investors. This activity is known as nation building, and the U.S. military does not like doing it, for it is expensive and troublesome and remarkably unmilitary. America lacks rich, prosperous enemies. Its worst enemies are either poverty-stricken or heading that way as fast as they can mismanage.

The American military is large and impressive, but its general trend is to shrink. It gets more powerful and danger-ous as it shrinks, because the advantages it has are technical, while the force of numbers merely slows it down. The core American military advantage, the linchpin of the New World Order, is information management. It is satellite and

aircraft surveillance, air and missile power, precision targeting, and rapidly deployed combined-arms coordination.

America has plenty of practice in exercising its technical
advantages. In the 1990s the U.S.A. fought major campaigns
against Iraq and Serbia. Its troops occupied Haiti, Somalia,
Bosnia, and Kosovo. The U.S. sent cruise missiles into Sudan
and Afghanistan. In the last quarter of 2001, with the Pentagon in flames, it went into particularly high gear. The U.S.
military is a consistently busy outfit. It's very likely to stay
busy dealing with the paramilitary consequences of its own
military success. If war is "politics by other means," the U.S.
military is facing unconventional struggles by means other
than war.

American military-design theorists have formally abolished the traditional battlefield. According to the doctrine of
their so-called revolution in military affairs, it's been rede-
fined as "the battlespace."

That sounds like mere high-tech jargon, unless you are
inside of one. Then the American battlespace is still somewhat abstract—to the American Nintendo warriors who are
inflicting it on you. But for the wretches on the receiving
end, it is a terrible reality. Short of an outright nuclear in-
ferno, American battlespace is the deadliest military arena of
all time.

It is sheer military folly to put on a uniform, formally declare war, raise a battle flag, assemble troops, and expose yourself to the digital targeting screens of American satellites,

cruise missiles, and aircraft. No conventional military force can enter American battlespace and survive for more than a few hours. This has been practically demonstrated repeatedly. In the Gulf War, American battlespace dominance was so complete that the fourth-largest land army in the world couldn't get anywhere near an American.

The NATO-Serbian war of 1999 was an even greater advance in this direction of robotic detachment. That time there was no battlefront at all. The NATO side suffered not a single casualty. This war was a one-sided two-month saga of American robot bombs destroying Serbian inanimate objects. As NATO bombs smashed Serbian bridges and power plants, NATO spokesmen would periodically apologize for unintended civilian casualties. As for the Serbians, their primary military response to NATO was to put on fresh T-shirts with target logos and stand on their bridges to sing.

The hero of the war that brought down the Taliban was the UCAV, the unmanned combat air vehicle. It's run by an intelligence agency and launches Hellfire missiles.

The political problem with battlespace is that it has no formal boundaries. Civilians may get apologies, but they have no space in which to retreat from the combatants. War goes undeclared, and so does peace. The Pentagon can become battlespace; so can the World Trade Center. Civil society is forced to militarize, while uniformed soldiers are safer than women and children.

There was gruesome chaos during the NATO-Serbian

war. It was not among the military, however. NATO forces took no casualties, while the Serbian military hid from the aircraft and was not badly hurt. Even though NATO was lavishly bombing the enemy capital of Belgrade, causing blackouts, traffic jams, and water shortages, there was no chaos there. Throughout the war, people of all ethnic persuasions were fleeing *into* Belgrade, eager to risk the bombing. Because NATO's bombing was much less dangerous and scary than the marauders on the ground. These men were not uniformed soldiers in any formal army. They were unconventional forces: paramilitary, lightly armed gangster guerrillas. These people represent the future of armed conflict. They started the war; they had the initiative; they set the pace of events. It was they who carried out ethnic cleansing, who emptied Kosovo of its majority population.

Bearded, swearing, sudden, and quick in quarrel, they swaggered from house to house and street to street, wreaking havoc with beatings, arson, explosives, and efficient, premeditated looting. Under the spacy oversight of American satellites and NATO high-altitude bombers, these bearded thugs, stinking drunk and screaming imprecations, were reducing civilization to ruins.

Without such men, there would have been no war in Serbia. Nor would there have been war in Chechnya, or Somalia, or Sudan, or Afghanistan. It is no uncommon thing for men to be veterans of all four of those conflicts, with a bonus trip to galvanized cells in Cuba. They are freebooters, jumping from nation to wreck of nation in the narco-arms bazaar.

Modern warfare is not all Stealth fighters, software, and Kevlar. It also has the dramatic aspects of a Shakespearean revival. Outside (and sometimes within) the prosperous bounds of the New World Order is a large and miserable New World Disorder. It includes not only the smoking ground of the Balkans but the Caucasus, South Central Asia, and vast, astonishing swaths of Africa.

Conventional war has become rare and brief today. But unconventional war is persistent and endemic. Great feats of terror are possible anywhere on Earth, while the rattle of small-arms fire and the periodic detonation of truck bombs are common around the world. This kind of war kills far more people than the high-tech Pentagon does. Even inhabitants of the Pentagon are killed by it. The heroes of these wars are swaggering condottieri, murderous, prickly, trigger-happy, and eager to avenge any slight.

To understand what this means and what it prefigures for the twenty-first century, let's consider three representative military figures who fit Shakespeare's description almost exactly. They are Shamil Basaev, Zeljko Raznatovic, and Abdullah Catli.

These three swaggering heroes are not household names in America. They're certainly not as famous as Osama bin Ladin, whose schemes made him the figurehead of visionary jihad futurism. They lack the grand torrent of global publicity that bin Laden brought down on himself.

A terrorist prophet leapt from the underworld onto the world stage in a spectacular Grand Guignol. That's unusual. Most garden-variety terrorist warlords are much more worldly than an aristocratic zealot haunting caves in Afghanistan. The pros of New World Disorder don't found any revolutionary schools of Taliban indoctrination. They know that it's hard to keep that level of fervor up and running. A suicide bomber has no encore.

Worse yet, a selfless world revolution by religious martyrs just doesn't pay. The Taliban were theocrats, but they ran their daily affairs through heroin smuggling, not through the collection plate at the mosque.

A terrorist mastermind is as rare and marvelous a thing as a criminal mastermind. Very few people engage in criminal acts because they are inherently procrime. They work at it because crime pays.

For the typical New World Disorder soldier, ethnicity and religion are not something you die for—they are stalking horses; useful pretexts for breaking down states and subverting police and governments. The resultant chaos can be structured, made to pay. Revolutionary idealists sometimes begin this process, but once the disorder fully flowers, their doctrines just get in the way. They will generally be rubbed out by greedier, more practical subordinates.

This process of New World Disorder is impressively globalized. It starts with local resentments, but it works through offshore money. It is financed not by the suffering locals under the heels of the warlords but by the New World Order itself. Civilized states require some criminal services. Their populations have a great and abiding need for narcotics, refugee labor, red-light districts, money laundries, and so forth. These are dangerous and socially ruinous activities. So civilized states persecute them and, by fiat, export disorder into the weakest and least policed areas of the globe.

In practice, these lucrative crimes have ended up housed in territories that are commercially worthless due to religious and ethnic tensions. So in a globalized world, the nations that break and fail become crooks. The workaday warlord biz talks religion all the time, but religion's not the daily round. The business is very unholy; it's all about oil, narcotics, guns, women, glory, and loot.

The New World Disorder will be a lasting aspect of the new century. It cannot pass from the Earth through a "war on terror" by states. Bold acts of mass terror can be made more difficult, but the need for the disorder's goods and services is endemic. It can't go away unless people become saints (unlikely) or states somehow acknowledge, manage, and legalize all the scary temptations of guns, prostitution, drugs, tax evasion, and (in the future) software piracy and genetic stunts.

For men like Basaev, Raznatovic, and Catli, chaos is a business environment that they must create and survive within. As we will demonstrate here in stage four, that is the way of their world.

In their own realms—Chechnya, Serbia, and Kurdish Turkey—these three men were everything that Shakespearean heroes should be. They were bold, daring, vengeful role models who took no guff from anyone, especially their own side. They inspired fanatical devotion from their followers, and they remodeled their own societies to suit their interests. They were inventive, determined, and successful.

A heroic military life requires a high casualty rate. No risk, no glory. These were three very glorious men. Shamil Basaev is the only one still alive at this writing. Raznatovic and Catli have already been martyred. As for Basaev, he has been declared dead by his enemies five times already. His survival odds look increasingly grim. But since he is still alive at the moment of this writing, we'll do him the signal military honor of letting him lead the way.

Shamil Basaev was born in 1965 in Vedeno, a small, rather backward Soviet village in the mountains of the Caucasus. An enterprising lad, he left the backwoods for the bright lights of Moscow in 1987. Fluent in Russian, he attended college for a while, hoping to become a police detective. But the Soviet economy was collapsing, and higher education failed him. Instead, he joined many fellow Caucasian ethnics in the new black markets thronging the streets of Moscow. Clever and forward-thinking, Basaev specialized in selling computers.

Then came the coup of August 1991. In his grandest heroic gesture, Boris Yeltsin defied the plotters from the top of a tank. Basaev somehow obtained three hand grenades and leapt bravely to Yeltsin's defense. But while Russia had a revolution, the Caucasus disintegrated. A rogue Soviet general named Djokar Dudayev decided that his native Chechnya should strike out for full independence from the crumbling USSR. Chechen rebels tore down statues of Lenin and seized the TV station. Basaev was a Chechen patriot, so he left Moscow and went home to his people.

While Major General Dudayev ranted, strutted, and steamed, the college dropout Basaev found the battlefront. Thinking big from the beginning, he joined a tiny but grandiloquent militia called the Confederation of the People of the Caucasus.

The Caucasus is broken into thousands of gorges and valleys. The ancient mountains swarm with small regional ethnic and religious groups: Abazinians, Cherkess, Dagestanis, Darginians, Laks, North and South Ossetians, Karachais, Balkars, Ingush, Avars, et cetera, plus all their various clans, parties, mosques, churches, and separatist factions. Even the local players can't tell themselves apart without a scorecard. (To complete the confusion, only the alien Russians call Chechens "Chechen." Chechens proudly call themselves Ichkerians.)

In 1991 shooting broke out in the formerly Soviet state of Georgia, among Muslim ethnic rebels called the Abkhazians. The Abkhazians had been restive under the mighty Soviets, but they certainly weren't going to put up with any hassle by mere local Georgians. Being Chechen, Basaev had little reason to care much about either Georgians or Abkhazians. But he considered himself a multinational revolutionary of the "People of the Caucasus." So he rode to the sound of the guns.

Up to this point, Shamil Basaev had, apparently, never killed anybody. He is probably best understood as an idealistic college dropout with a minor in arms smuggling. However, the small but vicious Abkhazian struggle turned out to be a matriculation camp for a generation of warlords.

In the heat of guerrilla war, Basaev turned out to have stellar leadership qualities. He was bright, affable, fearless, and loyal to his men. Practicing what he preached about Caucasian solidarity, he married an Abkhazian woman, which gave him useful influence among Abkhazian clans.

In November 1991 Basaev and his small gang hijacked a Soviet airplane and had it flown to Ankara in Turkey. He committed this terrorist act in order to publicize Chechen independence and make an international play for Islamic solidarity. The Turks were indulgent about this, though they wouldn't have liked it much if a Kurd had done it. Basaev was lionized and allowed to hustle back to the growing ex-Soviet fray. He then took an eager part in the brutal Christian-Islamic feud between Armenia and Azerbaijan.

In the summer of 1994 Basaev took a fateful trip to Afghanistan for hands-on training with veteran mujahideen. There, in the nascent world of al-Qaeda, he made many useful future contacts. He also learned almost everything there was to know about Islamic warriors killing Russians.

Basaev carried the Afghan tradition of the 1980s into the 1990s. He was extremely gifted at mixing age-old mountain traditions of ambushes, feuding, and banditry with brand-new weapons and tactics, and he showed a particular genius for using civilian cell phones to coordinate shoulder-launched rockets. Here his early computer experience seems to have given him a real boost. People who dismissed him as a poverty-stricken hillbilly would soon find that Basaev had websites, video cams, many overseas fund-raisers, and, often, admiring retinues of foreign journalists.

Major General Dudayev's independent Chechnya was an unqualified disaster. As a civilian leader, Dudayev was given to flashy suits and endless ranting on his state television, but he was unable to govern. He shut down the country's parliament in a coup. The schools and hospitals went derelict, the buildings unheated, and the staff unpaid. Violent crime sky-rocketed. The national treasury vanished into the hands of embezzlers. Electricity became spotty. Sewage filled the streets of the capital.

The educated urbanites who had formerly run the economy fled from the epidemic of kidnappings, muggings, and extortion. The remaining Chechen people had no choice but to live off support from their own clans, or *teips*, which functioned much like Sicilian crime families. Many angry Chechens rose against Dudayev with guns in their hands, defying his authority and turning every backyard and car trunk into a little mafia fiefdom.

Much the same dizzying decline was taking place in all of Russia, but a great deal of what was happening there was being done by Chechens. With their homeland out of control, refugee Chechens flocked to Moscow, swelling the ranks of mafia gangs like the Ostankino mob, the Tsentralnaya racketeers, and the Avtomobilnaya goons. The "independent" nation of Chechnya had the most porous borders in the former USSR. So Chechnya quickly became a duty-free gateway for foreign goods, which were smuggled in, trucked to Russia, and retailed in flea markets.

In December 1994 the new Russian government decided to restore the situation. They "intervened" in Chechnya, much as the Soviet Union had in the 1968 rebellion of the Czechs: with an overwhelming show of armored force.

The first Chechen war of 1994–1996 has become a model study in warfare. This was not one of those detached, antiseptic, almost dreamlike conflicts that America produces and presents on CNN. It never got much airtime, but it was vicious, bloody, gruesome, and thoroughly shattering.

Even though the country's leader was an air force major general, Dudayev's Chechen air force was useless. It was quickly bombed to pieces on the ground. The Russian air force had little trouble demolishing the Chechen airfields, detonating bridges and major roads, and toppling the television tower in Grozny.

But then came the nightmare on the ground. The Russian ground troops were hastily assembled from different units. They were badly trained and poorly coordinated, with no central command. Their radios didn't interoperate. They were a feeble, postrevolutionary army, invading a province in abject collapse.

It took the Russians weeks to reach the capital. The tank columns had to stop along the way to chase off indignant civilian protesters. Aging tanks broke down or got stuck on muddy roads. The Russian troops were poorly supplied, badly clothed, and maneuvering in winter. They often stopped to ask for directions and to scare up food, liquor, and cigarettes. They also bought and sold things in the Chechen black markets, including their own weapons.

The Russians slowly advanced on the capital in three columns, from north, east, and west. The incompetent commanders of the east and west columns could not make the necessary rendezvous, so the north column meandered into Grozny all by itself. Grozny, a large city that basically no longer exists, was then an oil-refining nexus of 400,000 people, a hundred square miles of Soviet-style brick and concrete high-rises, thick with grimy chemical plants and heavy industries. Many locals in Grozny were Russians, Ukrainians, and Jews, the civilian backbone of the region's economy. Since they didn't consider the Russians an enemy army, they didn't bother to flee the city. They just sat tight to await events, hoping for the best.

It was expected that Dudayev and his cronies would take the broad hint and clear out to the south, which had been conveniently left open for their retreat. On the face of it, retreat would have been an excellent idea. About 1,000 Chechen men under arms were facing 23,800 Russian troops. The Russians had 80 tanks, 208 armored personnel carriers, and 182 pieces of artillery, as well as an air force that could pulverize anything that annoyed it.

The Chechen soldiers were aptly described by the Russians as "bandit formations." They were homemade militias with borrowed uniforms, stolen weapons, and grandiloquent names. President Dudayev had a personal "national guard" of about 250 men. The Chechen "tank regiment" had a dozen working tanks. The Chechen "artillery unit" had only 80 men and 30 guns. The regular Chechen army added maybe 500 soldiers to this ragtag resistance.

However, there was also Shamil Basaev. He had no official rank at all and was generally known as the field commander of something called the Abkhazian Battalion. These two hundred volunteers were Basaev's personal retinue of terrorist commandos.

Unlike Dudayev, Basaev was a modern guerrilla rather than a comic-opera general. Interestingly, both Dudayev and Basaev had been to Afghanistan. Dudayev had been in the Soviet air force, blowing the Afghans up. But Basaev had learned all his lessons from the winning side.

Black-haired and brown-eyed, five feet seven inches tall, the twenty-nine-year-old guerrilla commonly wore a sheepskin jacket, jeans, and cheap running shoes. When he wanted to publicly show his allegiance, he wore a black knitted hat with an Islamic green ribbon around it. Except for his ever-present cell phone and Kalashnikov rifle, Basaev basically looked like anybody.

The Russian goals were the usual ones for a successful coup: they were supposed to seize the railway station, the airport, the radio and TV. Their primary target was the Presidential Palace, where Dudayev was holed up. On New Year's Day, Russian bombers set fire to Grozny's oil refineries, so the city was shrouded in smoke and winter fog. Confusion reigned; even as the tanks clanked forward, there were Russian parliamentary deputies inside the Presidential Palace trying to arrange a last-minute cease-fire.

The first Russian column rolled into town, checked in to the railway station, and started asking locals for cigarettes. They met scattered sniping and reacted with sporadic shelling.

The Chechen regime refused to flee the capital. Instead, as fighting increased, they used the open southern corridor to bring in more volunteers.

Russian T72 tanks were designed for the wide-open blitzkriegs of World War II. When choked and confined in cities, they become awkward weapons, with bad visibility and poor short-range fire. As military terrain, the high-rise streets of Grozny were much like the narrow valleys of Afghanistan.

But unlike the antiquated Russian tank, the Russian-designed RGP-7 rocket-propelled grenade is very modern: cheap, simple, portable, and accurate.

Shamil Basaev carried the traditional Afghan ambush into modern urban terrain. His men crept into place, coordinating with cell phones and little Motorola walkie-talkies. At a given signal, they would attack the lead tank and the rear tank of a column, using focused salvos of rocket-propelled grenades. The burning wrecks made advance or retreat impossible. No matter how they thrashed and boomed, the surviving tanks could then be picked to pieces. Russian tanks had never been designed to fire straight up. This made it simple to rain death on them from the tops of apartment buildings.

On New Year's Day the first Russian tank columns were massacred. Some Russians scrambled free from their burning tanks and were hunted down in wild street-to-street, house-to-house pursuits by bandits toting cell phones and swords.

After that gruesome debacle, a new and particularly savage form of urban warfare broke out. On the bantamweight Chechen side, it was dominated by rockets, Kalashnikov rifles, portable mortars, and radio-controlled booby traps. The heavy-duty Russians countered with the Grad rocket battery, which vomits forth a massive hail of explosive missiles, and with massed artillery, which obliterated entire buildings at point-blank range. To scour snipers from rooftops, the Russians used antiaircraft weapons combined with giant searchlights.

It was a war of scorpions versus sledgehammers, carried on in a city full of civilians. Tens of thousands of civilians were killed. This was well-nigh inevitable, because the best Chechen fighters dressed as civilians. They drove to battle in civilian cars, then opened up car trunks and removed mortars and missiles.

The urban guerrillas tormented the invaders with sniping and rocket grenades and were answered with crushing barrages. The Chechen guerrillas quickly learned to fire on Russian soldiers from apartment buildings full of Russian ethnics, who would then be slaughtered by the return artillery fire. An even cleverer tactic was to fire from between two Russian units, who would then shell each other.

The walls, floors, and ceilings of Grozny's high-rises were cruddy and substandard, while an AK-74 Kalashnikov is a rugged, powerful rifle. Blind, nightmarish battles broke out. Invisible Chechens would fire up from basements through floors, down from attics through the ceilings. They coordinated their attacks with cell phones. When the Chechens retreated, Russian squads inside the buildings would continue killing one another, chewing their way through the walls. After these demoralizing episodes of friendly fire, the Russians much preferred to pull out and smash everything.

The Russians probably lost five thousand soldiers in the first battle of Grozny, but the stunned and terrified local civilians took the brunt of the casualties. The dead cluttered the streets, unburied and often savagely mutilated. Winter was cold, dirty, and septic; troops could not find clean water and succumbed in droves to diarrhea and hepatitis.

Grozny was not conquered. It was demolished. It was a large city, a hundred square miles, and it suffered well-nigh constant firefights, sniper attacks, and rocket and artillery barrages, without any safety zone or defined battlefront. With no one left to tend the infrastructure, the ruined city sagged and collapsed under its own weight, with sewers clogging, power failing, and toxic fires consuming whole blocks.

Eventually the capital was ground down to smoking rubble. The Russians were able to raise a victory flag over the obliterated Presidential Palace. But the Chechen guerrillas were rather better off without their so-called government. There was still money around to finance the fighting, for the territory of Chechnya was webbed with oil pipelines. Little outlaw refineries sprang up everywhere, selling tax-free diesel and gasoline to thirsty markets throughout the Caucasus and Russia.

Attempts to seal the border proved useless. The new Russian nation was suffering a catastrophic depression. The black market had a stronger call on people's loyalties than the so-called Russian Federation did. The border guards were corrupt. Anyone at a choke point between the ex-Soviet economy and the rest of the world had a sure gateway to riches. Basaev boasted that he ambled in and out of Russia at will, even at the height of the killing.

In November 1995 Basaev engaged in the world's first act of nuclear terrorism, when he or his agents buried boxes of harmless radioactive materials in downtown Moscow. Then they called the press, and terror reached new heights of cruelty.

Communications technologies played a crucial role in the war. Chechen propagandists used cassettes, leaflets, video-tapes, pirate radio, radio jammers, and the Internet. President Dudayev obtained a portable television station with which to harangue the populace and threaten nuclear terrorism. The phone lines were open between Russia and Chechnya, so the families of Russian servicemen would re-ceive horrific phone calls from Russian-speaking Chechens, naming names and threatening personal vengeance. For their part, the Russians mounted loudspeakers on their heli-copters, which roamed the countryside bellowing threats.

After the abandonment of Grozny came a countrywide "pacification campaign." This meant that it was the villages' turn to be leveled. In the summer of 1995 Basaev's own hometown of Vedeno was thoroughly bombed, with the loss of many of his clan relations. Roused to suicidal fury, Basaev

gathered together his most ferocious mujahideen for a kami-
kaze attack on Moscow itself.

Basaev's *smertniki* were do-or-die suicide commandos.
They disguised themselves as Russian mercenaries in charge
of a convoy of Russian war dead. They repainted a car as a
"police escort," then boarded a pair of military trucks. Then,
incredibly, the 148 armed raiders simply drove off toward
Moscow, bribing border guards as they went. No guard
arrested them or even inspected the trucks. The commandos
might very well have rolled all the way to Moscow and shot
up the Kremlin, much as Kashmiri suicide raiders shot up
the Indian parliament in December 2001. Unfortunately for
this plan, however, Basaev ran out of bribe money in the
Russian town of Budennovsk. Here his car and two trucks
were escorted to the local police station.

Upon arriving at the station, the jittery commandos
boiled out of the cramped trucks and shot the local police to
pieces. They then rolled through the amazed and horrified
town of Budennovsk in a blazing gang, just as they had done
in Grozny, jumping from strong point to strong point, laying
down covering fire. They shot thirty-five people dead in the
streets. In the ensuing firefight, the mujahideen took some
casualties, so they retreated with their own wounded to a
nearby hospital. They kidnapped civilians as they went,
yanking people from their homes. The Chechens then com-
mandeered and blockaded the hospital, jamming innocent
captives up in the windows to block police snipers.

With fifteen hundred victims, including doctors, patients, and pregnant women, Budennovsk was the largest hostage seizure of modern times. It made Basaev's reputation.

The Chechens quickly made good use of the hospital's phone system. They demanded a cease-fire, safe passage back to Chechnya, and formal peace negotiations with Dudayev. In the meantime, they discovered that some of their hostages were Russian military pilots from a local air base. At least they were able to get to grips with their airborne tormentors, so they gladly executed the pilots in cold blood. The Russian media flooded into town, swiftly turning the bold act of banditry into a national political crisis. Two ghastly attempts to storm the hospital failed, killing 140 civilians in various cross fires. Russian nerve broke.

Basaev and his men returned to Chechnya in a triumphal parade of six buses, including a refrigerated truck for their own dead. The convoy included many Russian hostages, volunteer and otherwise, including reporters and Duma members.

Basaev released all his hostages once back inside Chechnya, but peace talks failed and the war continued. Russia enjoyed a rare victory when Dudayev was killed. He died, very characteristically, while ranting into a satellite phone. Russian electronic warfare experts used the phone's own signal as a homing device, so that the rocket barrage landed directly on him.

Far from harming Chechen morale, however, this loss of their deranged leader merely opened up operations for warriors much more competent. There was no more trifling about any constitutional legalities or official government appointments. From now on, it was a people's war dominated by heroes and martyrs.

Shamil Basaev was now a battle-hardened warlord with international fame. It would be wrong to say that foreigners flocked to his cause. Many brave volunteers were eager to join him, but he never commanded more than two thousand men in the field. It takes bureaucracy and organization to manage a regular army: regular soldiers have to be fed, equipped, mobilized, and sheltered. Basaev simply could not do this. If he had owned a barracks or a depot, the Russians would have swiftly bombed it into rubble.

The few fighters who were allowed to join Basaev, however, were as good as the international Islamic revolution had to offer.

For instance, Khottab. It's a rare thing in the modern world for a major public figure to have no identity. Generally there are school records, or fingerprints, or a birth certificate, maybe even some DNA samples. Shamil Basaev's chief lieutenant, however, is literally nameless and unknown. He goes by the nom de guerre "Khottab," variously anglicized as Khattab, Al-Khattab, ibn-ul-Kattab, Hottab, and Khotab. (He's also known as "the Black Arab" and even

"One-Armed Ahmed," for he once lost some fingers in a grenade explosion.) Khottab is variously said to be a Jordanian, a Saudi, a Saudi-Jordanian, a Bedouin Saudi, an Arab Jordanian, and a Jordanian of Chechen ancestry. (There's no reason why Khottab couldn't be all of these things at once, or none of them.) Khottab speaks English, as well as Russian and Arabic. Rumor says that his wealthy parents once sent him to an American high school. Since the Soviet war in Afghanistan, there's been no lack of American adventurers among the Islamic mujahideen.

If Khottab were merely an international man of mystery, an Afghan-Arab jihad war veteran, and probably the world's most formidable freelance religious terrorist, that would be remarkable enough. But he also specializes in media relations. Or, as he prefers to style it, "the Jihad of the Media." Sporting long black ringlets and a dashing Islamic beret, Khottab is the Che Guevara of the Islamic revolution. His fighters are always armed with video cams. They periodically upload video reports to the Internet, as well as distributing them hand-to-hand through the flourishing video-piracy networks of the Arab world.

These videos of Khottab's, which are not hard to find on the Internet, are no tasteful CNN-style edutainment docudramas. No Nintendo war here. They don't even have the exquisite formal Arabic and scholarly air of a bin Laden propaganda video. Khottab's jihad videos are shot right in the thick of ambush, amid flaming half-tracks, the yammer of Kalashnikovs, and battle cries of "Allahu Akbar!" Though Khottab's gruesome acts of free expression are periodically chased off websites through diplomatic pressure—he is, after all, an internationally wanted terrorist—English-language versions are only a mouse click away. It's the New World Disorder in your living room.

In 2002 the Russian FSB, successor of the KGB, claimed to have killed Khottab with poisoned mail. He may be dead. Or he may not be.

In the summer of 1996 Basaev revealed to the world that the Russian occupation of the ruins of Grozny was a mere Potemkin conceit. Swarms of his fighters rolled, trucked, and walked into the wreckage of Grozny from all points of the compass. Once again they began blasting Russian tanks with gusto, inflicting despair and panic all out of proportion to their numbers. One cutoff Russian unit was so panicked that it seized a Grozny hospital, taking civilian Chechens hostage. This terrorist act by uniformed Russians was probably the worst Russian propaganda defeat of the war.

In late 1996, aware that the war was a quagmire, the Russians more or less sued for peace. The political mess over Chechen independence was diplomatically finessed. The ruined province was pretty much abandoned. Basaev, crowned with a glorious victory, now turned his hand to governing the country. He ran for president and lost. But the victorious candidate, Aslan Maskhadov, seemed to trust Basaev. He made him prime minister and put him in charge of public security. Basaev's job was to restore public order in Chechnya.

Public order needed plenty of restoring. Except for the Afghan Taliban movement, no government anywhere was prepared to recognize the Chechen "republic." The country was a physical wreck: roads were mined, bridges broken, factories burned down. Oil smuggling was rampant, along with arms dealing. Thanks to Afghan contacts, the country also had a new, thriving line in heroin.

Faced with the consequences of his victory, Basaev failed abjectly. It's one thing to be a romantic hero who charges tanks with rockets. To walk the beat with a nightstick was beyond Basaev's ability. In the months that followed, the stricken nation fell off the edge of the Earth. The country swarmed with thieving marauders like a basket of crabs.

Wealthy foreigners were recognized as money on the hoof, so journalists and aid workers become hostages. Things reached a nadir in Chechnya when six members of the International Red Cross were executed in their sleep.

Outside Chechnya, hustlers and pretenders with forged credentials ruined the reputation of the so-called government. Even pious fund-raisers for the holy jihad were practiced swindlers who lined their own pockets with oil money from gullible Arabs. After years of bitter war and earnest guerrilla training, everybody and his sister had learned how to blow up cars and buildings. Remote-controlled assassinations became as common as dirt.

In the winter of 1997 Basaev and Khottab roused their disheartened troops and invaded the neighboring country of Dagestan. Some say they wanted control of oil pipelines, some say they wanted a seaport, and others think Basaev still believed his old pan-Caucasian propaganda. But the Dagestanis, though they didn't like Russians, were certainly not to be liberated by mere Chechens. Dagestan was already swarming with thousands of Chechens, squalid refugees reduced to utter penury. The Dagestani populace reacted with anger, loathing, and gunfire, and the border raid failed.

The larger war with Russia resumed immediately. After the Dagestan aggression, though, the Chechens were more isolated than ever before. This was rather good news for Shamil Basaev. One of the world's lousiest policemen immediately became a popular guerrilla hero once again. Though he had failed at peace, he was still first in war and first in the hearts of his countrymen.

The late Zeljko Raznatovic will be forever known as Arkan, since the world press finds that much easier to spell. Arkan was born in 1952 in Yugoslavia, in the small Slovenian town of Brezice. He was an army brat, the son of an influential officer in the Yugoslav air force. Though he was good-looking and had a talent for languages, he was a thief and a soccer hooligan. He made things too hot for himself at home, so he fled the country. The economic prospects weren't looking good, anyway.

Though Yugoslavia's socialist economy was stronger than those of the hard-line communist countries mired in the Warsaw Pact, it couldn't compare to the booming, labor-hungry markets of Western Europe. Left behind by technical advances, the Yugoslavs were increasingly living off money mailed home by émigré "guest workers." But the many emigrants working outside Yugoslavia were no longer under the thumbs of Tito's secret police. This meant that offshore terrorism became a lively possibility. In the 1970s Croatian separatists were some of the most ambitious and violent terrorists in the world.

The Yugoslav security forces were not on good terms with NATO. Though they badly needed Western cash, they were in no position to ask favors for controlling their dissidents or their terrorists. Given the circumstances, it was cheaper and simpler to have the Croatian terrorists shot by plausibly deniable counterterrorists.

We'll probably never know how Arkan, a violent, thieving juvenile delinquent drifting through Europe and cheering soccer games, managed to get this job. His father did have friends in state security. Arkan was good at languages, and he was there on the ground and available. Born near the border of Croatia and close to its capital, Arkan knew how Croats talked and acted. There's no question that, as a wild kid with a prison record, he was thoroughly expendable. Last but not least, Arkan always sincerely enjoyed killing people.

Arkan also had a talent for bank robbery. As fellow gangster Goran Vukovic fondly reminisced later, "Of all of us, Arkan robbed the most banks; he walked into them almost like they were self-service stores. No one can quarrel with that fact about him. I don't know about politics, but as far as robbery is concerned, he was really unsurpassed."

His chosen alias, Arkan, meant "arcane"—secret, mysterious, and known only to the few. Arkan kept his mouth shut about his Yugoslav secret-police sponsors, but once he was in the West, he engaged in endless acts of psychopathic daring. Western authorities kept catching him red-handed. In Brussels in 1975 he got a stiff fifteen-year sentence for various holdups. Somehow he escaped Belgian captivity in less than four years. Holland found him guilty of three other armed robberies, but he escaped the Dutch slammer in 1981. Later that year he was wounded during a botched robbery in Germany. He escaped his German captivity by breaking out of the hospital. At a robbery trial in Sweden, he and an accomplice held up the courtroom at pistol point and scrambled to freedom out a window. He also skipped out on court proceedings for a murder in Italy.

In 1986 Arkan, now thirty-four, took the fateful step of relocating to Belgrade. He had become a sophisticated, Westernized, multilingual émigré, back from the bright lights to open ambitious new operations inside Yugoslavia.

Step one was to acquire a money-laundry front for his illicit operations. Arkan established a cake and ice-cream store, near the Belgrade stadium of the Red Star soccer team, his favorite sporting outfit.

He also started a protection racket with local nightclubs. In the mid-1980s Belgrade was considered the rockin'est, most happening city in the eastern bloc. The music scene was booming, with hipsters showing up from gloomy Poland, Hungary, and Czechoslovakia, eager for hot girls, bright lights, and a bass beat. These shoestring hustlers also did a lot of "subsistence tourism," the well-established Soviet-bloc practice of cramming your luggage with cool stuff that you can sell back home. The well-traveled, well-connected Arkan was just the kind of guy to deliver those goods.

With his bakery store as HQ, Arkan became the chairman of the Red Star fan club, the Warriors. These Belgrade soccer thugs loved a good brawl—especially if it involved putting the boot to traditional fan-club rivals, like the Croatians, Bosnians, and Kosovars. Soon Arkan was the hippest mob boss in 1980s Belgrade. He had no problem recruiting cronies. His core rackets of sports, sex, and rock and roll (plus offshore car theft and generalized smuggling) were far more attractive than any boring Belgrade day job on some crumbling assembly line.

As dissent grew in Yugoslavia, Arkan's hooligans were very loyal indeed to the central government and their secret police. The crop-headed soccer boys made a particular habit of beating up wimpy peaceniks and intellectual dissidents, thus saving regular cops the trouble and embarrassment.

By the turn of the decade, Yugoslavia was breaking up into ethnic chunks: Serbs, Croats, Bosnians, Slovenians. In October 1990 Arkan turned his football club into an armed militia. In a spooky ceremony at an Orthodox monastery, he swore in a crowd of his best nightclub bouncers as the Serbian Volunteer Guard. They soon became known as Arkan's Tigers.

In November 1990 this amateur army set off on its first bizarre exploit, which was to assassinate the newly elected president of Croatia. Arkan and some of his boys, bristling with weapons, were arrested near the Croat capital of Zagreb. Arkan was found guilty of "preparing and participating in an armed rebellion against the state and sovereignty of the Republic of Croatia." He was sentenced to twenty months in a Croat prison. True to form, he was somehow out on the streets again in a short seven months.

Balkan tensions continued to build. Near the border town of Vukovar, local Serbs were in the majority, and they counterrebelled against the breakaway Croats. Arkan and his Tigers took a starring role among the Serbian militia. Unlike the dazed, ragtag local Serbs, the big-city Tigers were fresh from Belgrade, heavily armed, and kitted out in spanking new jeeps, courtesy of the secret police.

The Serb militia couldn't conquer the town street by street against tough Croat resistance. So they were joined by friendly units of the regular Yugoslav army, which still owned the massive cold-war arsenals of the Yugoslav air force and artillery. The regular army surrounded Vukovar and, over three months, obliterated it. The wretched town was so lavishly blasted that it received more bombs and shells than all of Yugoslavia had suffered in all the awful years of World War II. Finally the last battered Croats surrendered. The survivors retreated to a hospital.

Seizing the brand-new day, the Tigers took charge of the 250 Croat sick and wounded prisoners. They hauled them out of town in buses, then massacred them, killing off the medical staff as a finishing touch. They dumped the dead into secret mass graves, which were dug up and documented years later by somber war-crimes tribunals.

The siege had been very severe, but this deliberate atrocity turned Vukovar into a flaming symbol of unappeasable hatred between Croats and Serbs. However, Arkan and the Tigers were just hitting their stride. Unlike most modern wars, the Yugoslav civil war had borders and front lines. But holding some boring trench and munching cold K rations was never the Tiger style. The Tigers were a militia of a strange new kind: a high-flying death squad of mafia burglars.

The Tigers would ride into areas where the ethnic shooting had not yet begun. They raided small towns, carrying computer-printed hit lists of prominent non-Serbian locals, compiled for them by the secret police. Surprised civilians were tortured, raped, and shot, and their homes were looted and burned. Though the Tigers sometimes managed to avoid full-scale public massacres, they never failed to steal. They specialized in foreign cash, gold, and jewelry. These private hoards were lucrative and handy targets, since nobody trusted Yugoslav banks or the Yugoslav dinar. The Tigers also managed chains of cargo trucks that removed household appliances, furniture, bicycles, stereos, and even carpets.

The Tigers excelled at polarizing acts of race hate: wrecking graveyards, burning churches, and the particularly unattractive Balkan tradition of gouging out the eyes of the dead. Wherever the Tigers left their jeep tracks, everyone without exception was terrified—especially the Serbs, who knew that these psychopathic provocations could only bring fantastic vengeance down on them from former friends and neighbors. And vengeance indeed came. It came in massive military reprisals like Croatia's ruthless Operation Storm of 1995, which "liberated" the wreck of Vukovar and chased 250,000 Serbs from their homes. Though Arkan excelled at starting wars, the Serbs never won one.

Thanks to their secret-police sponsors, the Tigers always remained, somehow, officially unmentionable. From the Serbian perspective, they were always armored in political Teflon. This continued even when the Tigers raided Bosnia with their own media coverage and publicly curb-stomped dead civilians in front of Western news cameras. When queried by angry diplomats, President Milosevic referred to Arkan as a "simple sweet-shop owner."

By the mid-1990s a shocked Europe was imposing comprehensive economic embargoes on the flaming wreck of Yugoslavia. Embargo criminalized the whole Serbian populace, erasing the distinction that had once existed between Arkan and law-abiding people. Embargo proved to be Arkan's single greatest career boon. He become a vital economic lifeline for a large, formerly prosperous, formerly advanced nation of ten million people.

Arkan linked up with offshore Yugo mafiosi throughout Europe, and as the nation sank into penury, his own wealth and influence soared in direct proportion. Granted, he wasn't the only mafia warlord in town. From early on, wanna-bes were copping Arkan's best moves. Dragoslav Bokan ran the White Eagles. The Chetniks of Vojislav Seselj were widely known for their fabulous acts of cruelty in detention camps. The City Breakers of Franko Simatovic wore a weird Serbo-Texan getup of terrorist ski masks and cowboy hats. Arkan's worst personal rival, however, was probably Marko Milosevic, the glamorous, disco-owning, snappy-dressing son of President Milosevic. The two youth-culture moguls competed for the awed loyalty of the *dieselashis*—as the local fuel smugglers were known. First because they smuggled diesel fuel. Second because Diesel clothing was their favorite designer brand.

Arkan became an archetypal Serbian "patriot-businessman." He illegally imported most everything, from diapers to cigarettes. He bought in to a chain of hotels, started casinos, and sponsored a Belgrade shopping center locally known as Arkansas. He would drift in and out of the killing fields on weekend theft-and-pillage atrocity raids, retailing the proceeds to grateful Serbs.

As the long Yugoslav wars dragged on, Arkan became rather settled and established a regular routine. He spent much of his time holding court in his favorite posh hotel, the Belgrade Intercontinental. Since the hotel hosted foreign journalists, life there was nowhere near so cramped as in the rest of Belgrade, which suffered blackouts, fuel shortages, and hyperinflation. When not chatting with the foreign media, Arkan would hole up in his landmark Belgrade mansion, a vast Spanish-style hacienda with Greek pillars and a glass elevator. Like other music-scene swingers, he periodically married and discarded wives, leaving a trail of seven children. Eventually he was indicted as an international war criminal, but so were his best friends; besides which, Arkan was already wanted in several Western countries for his plethora of earlier jailbreaks, so from a practical point of view this indictment made little difference to him.

Arkan founded and headed a political party, which was his private army by another name. He ran successfully for public office, being rousingly elected by the anxious Serbian minority in Kosovo (local Albanians were not allowed to vote). He bought a failing soccer club who became champions as soon as the referees fully understood who they were facing.

However, the crowning moment of Arkan's career was undoubtedly his highly publicized marriage to a pop diva. After his second divorce, he kindled a torrid romance with the nation's sexiest and best-known singing star, Svetlana "Ceca" Velickovic. In February 1995 Arkan and Ceca put on a lavish, nine-day wonder wedding, complete with magazine photo spreads, live-TV coverage, and a machine-gun-toting honor guard. It was a festive national occasion, the brightest moment that stricken Serbia had seen in many years. The twenty-one-year-old bride wore a dazzling gown, aptly adapted from Scarlett O'Hara's defiant finery in *Gone With the Wind*. Arkan, forty-two, outdid himself in a dashing militia uniform, including a cape, jackboots, and a giant golden Orthodox cross. The newlyweds rode off in a maroon stretch Jaguar with a gold grille and California license plates. The country went wild.

The real secret to Arkan's success is that he was not merely a vicious, depraved war criminal but also a hip impresario who knew what the people liked. Arkan had spotted top talent. Ceca sang in the odd local "turbo-folk" idiom, and her stage name (pronounced "Tsetsa") was a little hard for foreigners to master. But she was a perfectly authentic pop star: a vivacious, good-looking babe with a voice. Ceca was the role model of a whole generational class of young Belgrade women, the "sponsor girls," who had grown up in the black economy as models and gangster molls. By marrying the mighty Arkan, Ceca was landing the biggest fish in Serbia.

After his 1995 marriage, Arkan disbanded the Tigers and settled down to rampant profiteering in his casinos, hotels, and shopping centers. Ceca gave him the eighth and ninth of his acknowledged children. This period after the Dayton Peace Accords seems to have been the most stable of Arkan's remarkable life. Becoming Mr. Pop Star seemed to have finally mellowed the warlord out. (Though when a former boyfriend of Ceca's, a top mafioso named Shaban, was conspicuously rubbed out, nobody was much surprised.)

The Serbian nation doted on Arkan. He was certainly the best-known Serbian military figure worldwide, far better known and much more feared than any professional Serbian soldier. He was glamorous, famous, and rich. He employed at least three thousand people and was owed godfather-style favors by thousands of voters. He clearly enjoyed making TV appearances on talk shows to promote his wife's Serbian pop videos.

When Kosovo finally ruptured, Arkan was forty-seven, losing hair, and putting on weight. He stayed rather low during the NATO-Serbian war, merely blustering to the foreign press at the Intercontinental. Even without personally sacking Kosovo, however, Arkan had a lot on his mind. After four lost wars, a collapsed economy, and massive out-migration, Serbia was dominated by the ethnic-terror mafias. Unless he started another war on someone unsuspecting and naïve, Arkan had less and less to pillage. Belgrade gangsters were reduced to quarreling over the dwindling spoils of their darkening city.

Comrade after old comrade was whacked over racketeering turf, shot in their restaurants and cars. Slobodan "Mauser" Miljkovic bit the dust. The same went for "Kundak" Todorovic, "Koca" Kovacevic, and "Ugar" Ugarkovic. Arkan's favorite secret-police boss, Colonel Radovan "Badza" Stojicic, caught a hail of bullets in Belgrade's Mamma Mia café. The casualty list stretched over six years to some five hundred unsolved underworld killings. Belgrade had become "a small pool with many crocodiles." These self-appointed Serbian saviors had been murdered in their own town by their own side, many of them quite likely killed by Arkan himself. The carnivorous killing spree migrated from lowly soccer thugs right up the chain of command to high government officials. Serbia's minister of defense, Pavle Bulatovic, was shot dead in a soccer-club restaurant.

During the NATO-Serbian war, NATO made a deliberate point of blowing up many of Arkan's favorite business holdings. Arkan prudently moved into the Hyatt, where the foreign press provided him with a kind of impromptu hostage corps. Ceca, for her part, bravely led much of the public singing on Belgrade's endangered bridges. But the NATO bombing did not let up, and when it did, Arkan found he had lost his political constituency along with his Kosovo holdings. He was a leader without a country.

On January 15, 2000, Arkan, Ceca, their favorite body-guard, Manda, plus Ceca's sister and a secret-police buddy, were all relaxing at Arkan's usual haunt, the Intercontinental. Two men came into the lobby and emptied Heckler & Koch machine guns at the party. The bodyguard died, the cop friend was also killed, the sister-in-law was winged. Arkan, who'd learned to wear a bullet-proof vest, took a fatal bullet in the face. The killers somehow spared the lovely Ceca.

Arkan's posthumous career began immediately. Thousands of tearful mourners attended his splendid public funeral. Death doubled his popular appeal. Old enemies who had denounced him as crazy criminal scum immediately forgave him everything and publicly pronounced him a Serbian knight, martyr, and patriot. Tiger guards defended Arkan's gravesite around the clock, since rumors quickly spread that he'd been buried with a hoard of gold.

Many in Belgrade thought that the entire event was staged; surely Arkan wasn't really dead. He'd surely given death the slip, just as he did so many prisons in so many places, to go underground again, using yet another fake ID. Someday, when a suffering nation most needed him, the hero would return.

Abdullah Catli was little known until he died in a car wreck in November 1996. The mere word "spy" scarcely does this man justice. He was a glamorous, charismatic, jet-setting politician, businessman, heroin smuggler, counterterrorist, mafioso, multimillionaire, bon vivant, assassin, and hit man.

Abdullah Catli's motives, weapons, and tactics were very much like those of Shamil Basaev and Arkan. But Catli was different, because he was firmly on the side of the New World Order.

Catli was the NATO version of Arkan and Basaev. He was not from some minor breakaway province like Chechnya or Serbia. He came from Turkey, a major regional power of sixty million people, in a strong alliance with Europe and the United States. He was a government-sponsored, CIA-friendly agent who heroically worked undercover to war against terror. He was good at it, too; but that was just the beginning.

Catli's biography beggars belief. It's so divorced from the official version of Western democratic law and order that it resembles magic realism. His story has it all. Hit men. Movie stars. Gangsters' molls. Heroin smuggling. Phone taps. Prison breaks. Car wrecks. False passports. The pope. The Armenians, the Kurds, and the Azerbaijanis. The CIA. The P2 Masonic conspiracy in Italy. Fifteen hundred AK-47 rifles. Crooked cops. Casino moguls. Labor unions. Communist students. Torture. Murder. Bombings. And, finally, great patriotic success on the heels of a personal tragedy.

Like Shamil Basaev and Arkan, Abdullah Catli was a small-town boy. He was born in 1955 in Nevsehir, in the rugged heartland province of Cappadocia. As a teenager he fell under the spell of Turkey's Gray Wolves movement. The wolf (which also appears on the flag of Chechnya) has long been the symbol of "pan-Turkism," the idealistic conviction that all people who speak Turkic languages should become the citizens of a united ethnic empire.

Catli, who was good-looking, smart, and a natural leader, excelled in Gray Wolf politics and youth activism. He became chairman of the Gray Wolf chapter in the Turkish capital of Ankara.

During his college days in the 1970s, Catli came face-to-face with radical Turkish student leftists. These Marxists were more than Catli could stomach, so in 1978 seven student agitators of Catli's generation were blown apart by a bomb at Istanbul University. Catli was tried in absentia for this act of terror, and he was even found guilty. Somehow, the Turkish police never quite managed to take him into custody.

In fact, not only did the young Catli not go to prison for killing communists, but in November 1978 he masterminded a jailbreak that sprang twelve other Gray Wolves from prison.

Catli followed up this feat in 1979 by arranging the fateful escape of Mehmet Ali Agca, an ardent Gray Wolf who had murdered a left-wing newspaper editor. Catli broke Agca out of Turkish prison, gave him false papers and a pistol, and spirited him off to Bulgaria. Two years later, in Italy, in 1981, Agca used Catli's pistol to shoot the pope.

In the early 1980s Turkey, much like Yugoslavia, was very

troubled by violent, ruthless offshore terrorists. In Turkey's case the terrorists were Armenians. Much like Arkan versus the terrorist Croats, Catli became an offshore global hit man against the terrorist Armenians. But he didn't rob banks as Arkan did. Instead, Catli covered his expenses with heroin.

Catli enjoyed great success exterminating the Armenian Secret Army for the Liberation of Armenia. But the Swiss apprehended him in February 1982 with four kilos of Turkish heroin in his baggage. Somehow Catli managed to mysteriously escape Swiss custody.

Catli was arrested once again, this time in Paris, transporting yet more heroin. The French were less understanding than the Swiss. Catli spent six long years in the French penal system, until the French sent him back to the Swiss again. In 1990 the Swiss mysteriously let him escape. Catli spent the rest of his life as a wanted international fugitive. Just as with Arkan, this legal technicality never slowed him down much.

By 1992 the Armenian terrorists, being mostly dead, were much quieter, but the Kurds were in open revolt. They wanted their own nation of Kurdistan. Unfortunately, this imaginary Kurdish homeland covered large sections of Iraq, Iran, Syria, and especially Turkey. The Kurdish rebels found themselves methodically ambushed by Turkish central intelligence and its two rival death squads, the Special Warfare Department and the Special Forces Command.

These two organizations consisted of about a dozen guys each. They were led by high-ranking professional Turkish policemen but were manned by dashing, expendable, hard-living Gray Wolves, such as Abdullah Catli.

Catli and his brothers-in-arms charged into the fray against the Kurds. Though they excelled at killing Kurdish terrorists, they were even better at making money in their drugs and arms rackets, which helped them finance the nation's secret war.

Soon, however, the gnawing acids of disorder ate their way into the framework of the state. The patriotic tail could no longer wag the black-market dog. The drug money was just too big and too tempting. The Special Warfare Department and the Special Forces Command became commercial mafias first, spies as a distant second. Naturally, just like all mafias, they started getting jealous over the turf and the loot. They even took potshots at each other.

In 1992 a prominent Kurdish casino owner named Omer Kaan failed to pay Catli protection money. Omer Kaan was found in a Range Rover in Turkish Cyprus, missing his head. After that episode, the owners of Turkish casinos everywhere did pretty much anything that Abdullah Catli said. Casinos are a good place to pick up girls and get a properly shaken martini, but they are also excellent money laundries.

In 1995 assassins tried to kill Heydar Aliev, the president of Azerbaijan. Aliev had been trying to clean up the Turkish casino influence inside his newly independent country. The failed assassins were never captured, and it was said that Catli was among them.

In 1996 Catli moved from casinos into media. He boldly kidnapped a TV mogul named Mehmet Ali Yaprak and extorted a ransom of four million deutsche marks.

In July 1996 Catli killed a newspaper columnist.

Later that same July, Catli and three policemen walked up to a rich casino owner in Istanbul and emptied their rifles into him in broad daylight. The four assassins were briefly arrested, but no Turkish policeman dared to hold them. Then Catli and his friends and sponsors in Turkish intelligence took over the dead mogul's extensive holdings and divvied them up. Catli received the valuable casino in the Istanbul Sheraton. As his business affairs prospered, he also became a corporate officer in BOTAS, a multimillion-dollar Turkish pipeline firm.

Mind you, these are merely the Catli activities that were revealed in detail in Turkish parliamentary investigations and hearings. These reports were based on sworn public testimony and written up extensively in open and democratic parliamentary hearings by the Turkish government itself, who deserve every credit for revealing these dodgy activities of their intelligence services. So does the Turkish media. Not many nations have the courage and valor to do this sort of thing. In fact, whenever seriously pressed by terrorism, nations generally do the exact opposite. No tactic

seems too dirty under those conditions, and covert opera-
tors who were once spurned are given money, guns, and carte
blanche.

There also exists an extensive dossier of legendary Catli
acts of derring-do that were *not* revealed or confirmed by
the Turkish government. These bits of unconfirmed gossip
include nuclear smuggling, aiding and abetting the jet escape of a criminal British billionaire, hijacking a Norwegian
cruise liner, bomb attacks on leftists in France, and setting a
Greek island on fire. Perhaps Abdullah Catli did these amazing things; perhaps they were done by his colleagues; perhaps some were never done at all. But Catli had the firm
reputation for doing them, and while he lived, nobody dared
to cross him. He was a smack-dealing terrorist with a government badge, and his friend had shot a pope. For those
who knew of him, he was truly one of the scariest men in the
world.

In November 1996 Catli was enjoying a high-speed spin
in a posh Mercedes driven by his good friend the chief of the
Turkish police academy. But Turkish traffic is tricky, and the
Mercedes ran into a truck. The driver was killed, and so were
Catli and his latest and most glamorous girlfriend, Gonca
Us, "Miss Turkish Cinema."

Traffic cops showed up at the smoking wreck and found
that the late Mr. Catli was carrying cocaine, a handgun, and
a wide variety of faked passports. Even worse, the trunk of
the crashed Mercedes was crammed full of illegal guns.

Vague attempts were made to hush everything up, but Catli's underworld was just too big to hide. Even his secret sponsors had lost track of what the guy was up to. Thus erupted the "Susurluk scandal," the worst and biggest political scandal of Turkey's twentieth century.

The resultant Watergate-style investigation brought down the Turkish government. There was great indignation among the Turkish people when they learned that a puffed-up superspy had been "walking around like a minister," gunning down his victims in public, and laundering millions of dollars in dope and casino black money.

But Turkish governments, like the Turkish people, are very resilient. Slowly the awe and the wonder faded, and the Turks just patched together another government. With the passage of time, Turkish sentiment toward Catli seemed to change. His rampant corruption was forgiven, while his successes rather spoke for themselves.

If a war on terror is "a new kind of war for the twenty-first century," it's quite old hat for the Turks. They have a conspicuous, if wondrously dirty, record of real-world success in this arena. Turkey is the world's most successful and stable Muslim state, a Muslim economic superpower, and compared with their enemies and neighbors (all their neighbors are their enemies), the Turks look great. The one Muslim state treated as an equal by the West, Turkey is an honored ally and not a client state of anyone.

Turkey's enemies have not done at all well. Many dead

Armenians, many dead Kurds. Wretched Serbs, bankrupt Syrians, stunned Greeks, bombed communists. Turkey's bête noire terrorist mastermind, Abdullah Ocallan, was successfully chased from nation to nation, denied any safe haven, and finally abducted in Kenya. He was carried back to Turkey in chains, to the huge satisfaction of the Turkish voters.

Even wretched, humbled Russians ended up haunting the streets of Istanbul, looking for bargains to stuff in their cardboard suitcases. Across Central Asia, a huge swath of Turkish-speaking peoples looks up to Turkey as the model of a way forward.

When it came to global trouble with terrorists in Afghanistan, the Turks were the first to volunteer to sort things out on the ground. If Abdullah Catli were alive today, he'd be precisely the sort of dodgy figure whose help would be ardently sought. And he would exact the same price, for those who fight the New World Disorder on its own terms and on its own turf come to wear its face and its clothes.

Catli's grave is honored today, and Turkish street protestors in Paris have been known to carry picket signs reading CATLI LIVES.

In concluding Stage Four: The Soldier, we may now ask ourselves, in the jargon of the American military: what are the "lessons learned"?

The three men we've studied have remarkable commonalities. First and foremost: they're not real soldiers. These marauders wreak serious havoc, but they have none of the traditional military virtues. They lack obedience, discipline, order, long martial traditions, formal esprit de corps, and most of all, the ability to establish and maintain a national state.

Order is hard work. Nation-states shudder and collapse when men like this show up. They live outside the law and they cannot keep order. Even if they succeed in avenging themselves and doing grievous harm to their enemies, their governments fail or fall. They may become a secret source of national pride, but they generally end up taking ferries in the dark to the husk of a burnt-out hospital.

All three men specialized in protection rackets. Rackets may look like taxes, but let's be clear: a government demands taxes in exchange for some range of social services. A protection racket demands money to leave you alone.

Two conditions are necessary for a protection racket to flourish. The racketeers must be obvious to the population (since no one pays protection money to someone they've never heard of) yet also invisible to the government. The end of a careful public silence means the death of a protection racket. So silence must be engineered and maintained. Not just any marauder can manage to be famously obscure, notoriously secret, and glamorously unspeakable. One lousy car wreck or outspoken photographer, and the whole thing can burst with an almighty stench.

Arkan, Basaev, and Catli were famous men whose governments, media, police, and military routinely denied their activities. They all existed in formally established, fully deniable demimondes. These half-worlds are very special places in society, where men like these have great freedom of action yet never face consequences.

They needed armed power, yet they never commanded real armies. So they were very inventive at arming the people at hand. They used football fans, black marketeers, mafiosi, smugglers, nightclub bouncers, political parties, youth camps, mountain villagers, mosques, churches, cops, spies—almost any organization that could be shoved over the line from civil society to the paramilitary.

These men claimed to be superpatriots, but they never spent much time with their own people. In fact, they seemed to find their own people rather tedious, parochial, and boring. Though they made a big deal of their blood-and-soil folk heroism, all of them spent their formative years as polyglot exiles. They fleeced and exploited the foreigners in richer nations. Back at home, they remained globalist glory hogs, always eager to make a powerful impression on the powerful people outside their own countries. They cultivated the foreign press; their most trusted confidants, co-conspirators, and mentors were often foreigners. They owed much of their success to generous foreign backers, and to their own ethnic diaspora: to Islamic mujahideen, to Serbian mafia in Europe, to offshore Turks, to people who could feed, shelter, and supply them without having to actually touch them.

Their physical courage should not be questioned: no schoolyard bullies, they were tough, ruthless, and careless of life. But they were always hit-and-run guerrillas who never stood their ground. Nothing was commoner for these men than to show up in a blaze of glory, provoke holy hell with some thoroughly calculated atrocity, and then scram, leaving their local fellow ethnics to bear the consequences of their actions. If they ever protected, sheltered, or defended any innocent civilians, it was by accident. Two of them ended up with death raining from the heavens on their own homes. The third ruined his sponsors and destroyed their agencies.

They all began with back-alley shooting and ended up in economics. This is the natural career arc of a modern war-lord. Sooner or later he comes to realize that true domination is about the money and not about the guns. Because money will get you guns much more easily than guns will ever get you money. Some dealt drugs, some dealt arms, but all three of them dealt in oil. Drugs are toxic, and arms often more trouble than they're worth. But no modern society can exist without oil.

In the world's black markets, the *dieselashi* is king. Even the squeamish Americans will climb out of their Stealth planes and risk a bloody land war for the black gold of the oil fields. For both the New World Order and the New World Disorder, oil is the number one source of global insecurity.

Without question, oil is the most dangerous contraband in the world.

We can't let these men depart the stage without a round of applause for the police. These three men all had a strange, symbiotic relationship with the police. A functional police force will not suffer the existence of people like this. But a dysfunctional police force *becomes* people like this.

Shamil Basaev went to school to become a policeman, and he failed. He later became the top cop in his stricken country and failed again. Arkan and Catli were secret policemen by trade, intimates of the top law-enforcement people in their own governments. They paid them off with proceeds of crime and thoroughly corrupted them. Arkan and Catli were state-supported killers with multiple legal identities and numerous aliases, aided and abetted by law enforcement. They both found prisons to be strangely porous.

Perhaps we can sum it up. These are the ABC's for future Arkans, Basaevs, and Catlis.

1. Male 25–40, likes handguns and has visceral, hands-on experience with face-to-face violence.
2. Extensive prison history somehow enhances his public reputation.
3. Quite good-looking, enjoys making a show in posh casinos and hotels with a line in feminine arm candy.
4. Speaks several languages, has spent much time in other countries.

5. Fluent, fast on his feet, media-savvy, good on TV.

6. Has a personal posse of devoted tough guys who are known by their nicknames—or, better yet, by the fake IDs that they got from cops.

7. Gets richer and more influential as life gets harder for his neighbors.

8. Emits chauvinist rhetoric but kills many people of his own nationality. Killing rivals in gangland a particular specialty.

9. Lousy at straight jobs. Can't take orders. Essentially unemployable.

10. Prominent in politics, eminently electable, but has no political philosophy, no sensible platform for governance, and no legislative or executive experience.

People with these qualities exist everywhere, in all cultures, at all times and places. A cruise missile, when you think of it, is just a rich guy's truck bomb. The risk to the future happiness of mankind is in the movement of a slider bar from local hell to general war.

We're fortunate that complex weapons dominate the battlespace, because complex weapons can be created and maintained only by a complex society. Though Basaev deftly uses "the enemy's weapons," he's not inventing any of them. He has no industrial base, no military-scientific R&D. He can light bonfires inside the New World Order and perhaps devastate wide areas, but he cannot build anything. He can't even defend the patch of ground that he himself is standing on.

Terror can destabilize, it can drive order away, but terror can't govern.

The New World Disorder offers so little to its victims (and even to its supporters) that it has always been in constant motion, like an epidemic. However, the disorder might perpetrate itself indefinitely if it could find or create some septic physical area so thoroughly ruined that order can never penetrate. The disorder is therefore unique among political arrangements in that it can thrive after nuclear, biological, or chemical attack. A Pakistan devastated by nuclear war, for instance, would become a super-Afghanistan: lawless, deadly, poisoned, and unpoliceable, yet still profitable to brigands, still a primal source of drugs, guns, gangs, and enslaved women.

Worse yet, the New World Order is so physically dominated by Washington that the city of Washington, D.C., is its single point of failure. Washington is certainly the world's best, most tempting target for a mass-destruction weapon. Should Washington be destroyed, all other centers of political authority will be in extreme danger. Cities and civilization may be physically liquidated in a ferocious Pol Pot style. Should that come to pass, the disorder will reach its apotheosis. It will reveal itself as the precursor of a planetary Dark Age.

The best defense against New World Disorder is not brave men careless of their own lives, charging the mouths of cannons. It is decent government and a functional economy, stubbornly maintained despite all provocation. The New World Order dominates by hardware support. Stable government underwrites that process, giving boring states with good functionaries the edge over that heroic chaos in the badlands—so far.

If you want order and decency, strange oaths and sudden quarrels won't do the job. This is the realm of a very different kind of human being: someone fit to govern. Shakespeare called him the Justice.

STAGE 5 THE JUSTICE

> *And then the justice,*
> *In fair round belly with good capon lin'd,*
> *With eyes severe, and beard of formal cut,*
> *Full of wise saws and modern instances;*
> *And so he plays his part.*

Contemporary politics comes in three basic flavors: Technocratic, Nostalgic Activist, and Bizarre.

Technocratic is the chosen flavor of the successful leaders of the most successful societies. Consider George Bush I, Al Gore, Gerhard Schroeder, Tony Blair, Lionel Jospin, and every Japanese politician ever elected. These are colorless, postideological technocrats who are devoid of grand political vision. They may well be forced into defensive crusades by weird situations that they never planned, but they don't

march to their conference tables with any grand reforming
schemes.

The Nostalgic Activists offer leaders with rather more
color, such as the very colorful Newt Gingrich and the involuntarily colorful Bill Clinton. They got plenty of airtime and
column inches, but their efforts rarely amounted to much.
Both of them were almost ridden out of their capital on rails.

The edges of the New World Order are flaky. Here we
find various untoward spectacles: French fascisti, wanna-be-
president computer moguls, and the "summit hoppers" of
"global civil society"—contemporary Reds, Blacks, and
Greens, who specialize in confrontational street theater. In
reaction against the dominant technocracy, the people of the
New World Order dote on political fringe cults, not from
any serious discontent with their lot but for the entertainment value.

The New World Disorder offers warlords and cult leaders who are always bizarre and often deadly. This is the
unhappy realm of Balkan warlords and al-Qaeda.

The grand global trend is crafted political blandness
with swift, unpredictable outbursts of wackiness and scandal. It is rather often punctuated by spasms of terror and
global emergency. These scary crises have not lasted, but
they do seem to be an inherent part of the system. The New
World Order got its baptismal name when the Gulf War
took place and was best understood as the sensible part of
the planet, the technocrats with the plumbing, the people

who were not on fire. Later, the Serbian war did a great deal
to firm up the security arrangements within and outside
NATO. After Serbia, it was pretty clear who would do the
shooting and who would nod and pay up. The "War on Terror" is more of the same: horrific and aberrant wherever the
buildings and bombs fall, but primarily diplomatic and economic. Every savage jolt is another opportunity to shore up
and rationalize the rickety international order. The wreck of
Afghanistan burns, but at the same time and for the same
reasons, China joins the WTO, India joins the inner military
circle, and Russia's defeats are avenged by its former foes.

At the global level, emergencies weld nations into un-
gainly grand coalitions. It makes sense, for the nations' governments lack real political divisions. Even in Europe—always
known for taking its doctrines with grand philosophical
seriousness—the distinctions in political philosophy are
mostly traditional gestures. The real gap is no longer between ideologies but between people who need to govern
and people who need to grandstand.

Political activism is struggling to delaminate itself from
government. Ideology longs to become something grand
and aspirational and wondrously impractical. In the twenty-
first century this separation might even become institution-
alized, the way the eighteenth century formally separated
church and state.

Do political activists need a government for their quarrels and campaigns? More and more, the "nongovernmental

organization" (NGO) looks like their natural milieu. American warplanes blew up Kabul. They hit not just the international al-Qaeda terror group but also the International Red Cross, which was already there in the very thick of the armed disorder, shuffling food and bandages.

Within the U.S.A., the right and left fuss about family values and corporate greed. The right wing wants to leave the market alone but to regulate sex. The left wing is relatively tolerant of domestic license but wants to regulate private industry. So, unless the state is restored to health by military emergency, this would seem to leave government a golden opportunity to do nothing.

Whenever the ideologues of right and left are out of power, they go right on agitating in the cultural sphere, often with undisguised new energy and a clear sense of evident relief. Then the right wing's effort becomes "culture war," which is carried out in parapolitical arenas like pulpits and schoolbook approval committees. The left-wing agenda becomes "socially responsible investing," which is carried out through boycotts, labor strikes, and product-liability suits. These ideological enterprises don't need government to achieve their ends. It would vastly simplify the drudge work of effective public service if zealots stopped dragging the technocrats into their crusades.

This withering away of the glamorous and exciting organs of the state leaves "government" a neutered enterprise, stripped of all heroism, courage, romance, and charisma.

But government is all the more functional and even successful thanks to its near-total loss of hope and aspiration. Government still has plenty to do; it's just that it's all very
boring. In this utopia of tedium, public affairs are properly
left to policy wonks who never rabble-rouse. They concern
themselves with banking laws, commercial laws, contract
laws, business codes of conduct, property rights, accounting
standards, officeholder ethics, tax laws, and insider trading.
Assisted and pestered by lobbyists and pressure groups, they
act as level-playing-field referees for the panoply of for-
profit enterprises.

These public servants have a clear goal, even a kind of
mission statement. But there is nothing idealistic about it,
no New Soviet Man, no Vision Thing, no American Dream,
no Holy Caliphate. There are no wars of conquest, no revo-
lutions, no Great Leaps Forward. Their political model is a
stable capitalist nation-state, with transparent accounting,
low inflation, price stability, a small public sector, a balanced
budget, low tariffs, open capital markets, a politically independent central bank, direct foreign ownership of industrial
assets, a fully convertible currency, privatized telecommunications, and, in fact, the privatization of pretty much every
enterprise privatizable.

That's the story. That is the management objective.
There's very little in the way of atrocities, pogroms, cam-
paigning, marching, rioting, or even much interest in voting
in a country like this. It's like Liechtenstein, only less excit-

ing. It's pretty much the way the world's smallest, humblest states have always behaved. Polite. Predictable. Ingratiating.

For a great nation, this situation is awfully demeaning. No patriot could desire this squalid destiny for his empire. But it's happening anyway, and for a very good reason. This kind of modest good behavior is what *foreigners* want from your country. That is why major governments find themselves behaving like minor ones. Their freedom of action has been severely reduced by swift international flows of money and media.

In the globalizing New World Order, foreigners are your major markets. They have the power to swiftly wreck your currency, to destabilize your banks overnight. Thanks to telecommunications, everybody gets a look at how you run your affairs. In the twilight of sovereignty, all nations behave like small nations.

Since they are smart and rich, those foreign investors are entirely indifferent to your phony-baloney national mythology. They could care less about your manifest destiny and your special character as a chosen people. They regard your patriotic sentiment as a symptom of megalomania. They may feel very ardent about their own country, but they won't tolerate any pretension from yours. They just want to know if they can bring capital investment into your region and get a good return on it. They will yank that money out at the speed of light if you show any signs of zealotry.

There are zealots galore, but there are no First, Second, and Third World Trade Organizations. There is only one. There's no workable method in which a nation can prosper alone. If you darken the network at your national borders, it means swift industrial decline and the emigration of your best and brightest. Where are the bright people emigrating; where are the best brains draining? Where do the planet's skilled and wealthy diasporas choose to live? Put a pin in the map there; tomorrow those places will thrive.

The New World Order conspicuously lacks a New World Civil Society. National politicians don't waste much energy on the nonvoters outside their doorsteps, unless there are screaming, flaming international emergencies, in which case nations fly in bearing arms. The New World Order still depends on nation-states to supply its civility, even as nations dwindle, disorder erupts, and postgovernmental networks grow stronger and often crazier.

Trade issues have ranked first and foremost in globalization—because that is the simple part. Great progress has been made negotiating the shipment of goods and services across national borders. The New World Order is a highly materialistic regime in which nothing but cargo is allowed to be serious business. Many people find this situation politically stultifying, intellectually barren, commercially exploitative, and morally scarifying. But analyzing that situation is not the same as solving it. There is no functional alternative, no working model for a different global way. There are no credible ideas of how to create one.

The bipolar struggle between state control and state capitalism ended with the cold war. Today there is no such competition of ideas. There is just functionality versus dysfunctionality, technocratic versus bizarre. You compete in the global market or you can't. Some countries go fast, some go reluctantly, but there's only one direction to go. Fail or refuse to compete, and your ultimate fate is clear. It's the gutter. The red-light district. The tar pit of the New World Disorder, which we visited in stage four. It's that vast pariah realm of warlord narcoterror.

Political ideas don't matter, but government still does. Anyone who's ever stepped out the doors of a Western-style hotel into the streets of a developing nation can see how much government matters. One can see stark physical differences in the planet's landscape just by looking out an aircraft window at the borderlines of counties, states, and nations. There's still plenty for governments to do. The planet's abuzz with potent activity. But it's not ideological. It's centered on performance, function, and return on investment.

Technocrats are dominant, not because they have the One True Way that is the culmination of history but because they support technical invention. They are willing to abandon the moral certainties of previous centuries and expose their populations to radical levels of postindustrial instability. They dominate competitively because they are willing to rapidly alter the means of production and provide wider opportunities. Technocrats don't mind risking the toxic side effects of network TV, traffic jams, nuclear weapons, and gene-warped food; of monoculture crops, dot-com crashes, BSE in the burgers, and HIV in the blood banks. They endure these downsides because there's a whole lot of money in the upsides, and the guys with alternative concepts didn't pan out. It was worth it.

The New World Disorder is a place of abject failure. The wretches there are desperately reckless, but they cannot prosper by their risk taking, because they are too unstable to support research and development. So they become the global slum, brothel, and opium den, earning their money from the vices of success.

Rigid authoritarian states—the grand experiments in defunct models—fell out of fashion. They could launch outer-space media spectaculars, but they couldn't manage innovation in everyday life. Schumpeterian-capitalist "creative destruction" outdistanced the Permanent Revolution. The places that create innovation have dull, cautious, shrunken, functional governments.

Once upon a time, technocrats were scarce. Before World War II, governments were notably reluctant to support scientific and technical research. Longhairs and eggheads cost a lot to feed and shelter. Their abstruse products were of dubious relevance to politicians. Then came radar and the Manhattan Project. The cold war began with a bang.

The stunning success of Vannevar Bush's Manhattan Project made the U.S.A. the first atomic superpower. This created a new and clear political understanding between Washington and technocracy. In the war's aftermath, Vannevar Bush founded the National Science Foundation. And the business of the National Science Foundation was National Science.

American scientists would receive national funding and supply high-tech preeminence to America. This meant preeminence not just in nuclear physics but in every discipline underwritten by the taxpayer. American science would methodically explore the "endless frontier" of new knowledge, while government paid the bills. This was a brilliant plan of social reform, on the grandest possible scale. Like the Bomb, it began as the pipe dream of a small cabal of intellectuals. But like the Bomb, it worked. It changed everything.

Big Science ensued, creating the bedrock of today's technocracy. But there's a qualitative difference. The New World Order's science is no longer National Science. It's not restricted to nation-states, and it's no longer about knowledge. New World Order science is not about military security and the public good. It's about intellectual property. Not about command and control but mind share and market share. This leads to a new form of dominance that is not about obliterating people with atomic bombs but about obsolescing them with networks. The ideal foreigner today is someone who is peaceful and dependable yet forced to play your game. His country is always trailing the leader by a respectful technical distance.

A government under these conditions is still a government, but it's not a people-in-arms. It lacks romantic slogans and aggressive doctrines. It has no enemy states, only competitors. A government in these circumstances can no longer claim a manifest destiny or call itself the Last Best Hope of Mankind. Politicians still try this kind of rhetoric, for it's not as if they don't want the job; Tony Blair in particular seizes almost every opportunity to sound a Churchillian trumpet, so much so that Britons complain that their nation seems too small for such grandeur. But even American politicians, the titular leaders of a genuine hyperpower, sound rather silly when they strike grand themes of Infinite Justice and Ages of Universal Liberty. It sounds like the guy who runs the hardware store claiming to be Napoleon.

A government need not take any road to utopia to accomplish its work. Government can take a long view where industry cannot. Expand the networks, keep the roads and airports open. Favor open systems, which expand general opportunity, over closed ones, which make a quick killing for oligarchies and monopolies. Provide a safety net to allow the population to adapt to changing circumstances. Keep the courts and elections honest. Keep sewers running, keep plagues at bay. These things do not require massive commitment and mindful sacrifice from the aroused masses. They're not sacred, they're not romantic, they're hardly even patriotic. They're subtle acts of quiet statesmanship, but they must be done. When they're done well, people prosper, and where they're done badly or not at all, the people are wretched.

Shakespeare's justice, with his portly stride and well-cut beard, would be a very happy guy in the former kind of establishment. It may sound like the "end of history," but aspiring to supreme historical significance has a large market downside. From the perspective of the twentieth century, riddled with screaming fanatics and totalitarian death camps, this calm, mature, and rather vapid setup sounds just peachy. It's a civilization that is functional, practical, sensible, and maybe even beautiful, if they hire the right guys to do their set design.

For a better understanding of twenty-first-century politics, we have to find the factors within the period that have no historical precedent. These factors will lend twenty-first-century politics its distinctive native character. Where's the part that's true to the times? Where's the source of genuine novelty? It's not the humility and dullness that makes tomorrow's government distinctive. It's not even the suicidal desperation of the excluded and fanatical. It's the amazing level of network mediation.

Throughout history, communication has been an arcane, expensive, and difficult enterprise. Government ran it. Literati were a small fraction of the populace, paid to put

ink on paper and maintain the necessary records of grana-
ries and taxes. They were Egyptian scribes, Chinese scholar-
bureaucrats, medieval clerics.

The twenty-first century lacks those former limitations.
It is boomingly burdened with storage and bandwidth,
while literacy rates are sky-high. So tomorrow's New World
Order is by no means an everyday historical empire. It is a
military-entertainment complex riddled by networks with
pseudobiological properties.

In these conditions a national government can't define
official reality with a few gilt proclamations. In emergency
conditions, it will do its very best in the way of censorship, spectacle, and spin—thus the "military-entertainment
complex"—but outside its own patch of real estate, everyone else tends to scoff.

Given enough media hardware and good reasons to use
it, everybody becomes media. Media engulfs the works; even
the Luddite terrorists deliver video sermons for the Media
Jihad. Politicians find themselves merely one class of network users. They are not the ones controlling the global network traffic, either. They are merely the ones held legally
liable when everyone panics.

If you look at the digital network's major power players,
they're certainly not politicians. They are multinational
media giants like AOL Time Warner, monopolist code moguls
like Microsoft, and intelligence apparats such as ECHELON.

Mind you, these various entities didn't build the Internet; they can, however, make sure that nothing remotely democratic can succeed there. This first entity routes around governments, the second crushes them in court, and the third doesn't even officially exist.

If you examine contemporary media coverage of American politics, a very strange thing becomes obvious. Though all commentators routinely cite the "American people," nobody wants to actually *be* the American people. The American people are a kind of official mythos, like phlogiston or the philosopher's stone.

Every actual American in the United States considers himself a sophisticated courtier. Nobody ever takes a political speech at face value. Political statements are always closely analyzed by pundits for their calculated appeal to interest groups and demographics. So no American is ever taken in by these speeches. Instead, everybody's a spin doctor, tuning in to talk shows to hear what they're supposed to be told to believe. Politicians are praised for staying on-message and pursuing Machiavellian charm offensives. They are ruthlessly dismissed as crazy, stupid, or past it if they seem to be believing their own press releases.

Nobody goes door-to-door in precincts, rousing the inert population to political awareness. American politics has become a very hands-off, abstracted, virtualized process. Campaign managers raise large sums to buy ritual displays of media, which are then psychoanalyzed for their effectiveness by televised opinion makers. Simple ardent political conviction evaporates in these conditions, replaced not by mere cynicism but by a media-literate population who are sophisticated enough to think and behave like political operatives.

The business of political operatives is horse trading in smoke-filled rooms. Most political issues offer a public place to grandstand over principle, and an elite arena in which to split the difference. This isn't hypocrisy; this is management. With the passage or frustration of every bill, you tally up your profits and losses, and political life moves briskly right along, even though this necessary tidying up of unbearably complex problems may well crush some widows and orphans. It's like pop music, housework, and the tides: it's cyclical and inexorable. It may be rather sordid and sleazy, but there's something reassuringly human and timeless about it. The people who succeed at this work enjoy their game; they are professionals. They're media professionals; they buy media for their clients, and they haunt the talk shows as media stars. And it works.

Except, that is, for outbursts of the bizarre: scandal and terror. Sometimes everyday politics is disrupted by an advent so wicked and heinous, so beyond the pale, that it calls the whole system into question. Then the apparent corruption, cynicism, and corrosion of ideals suddenly becomes intolerable. Like infidelity in a marriage, it's a transgression so gross, so inflammatory, so entirely an affront that it cannot be reasoned away or papered over. Years of selfless public service and technocratic stability cannot make up for it. The order of the day is threats, recriminations, saber rattling, sobbing histrionics, and unanimous wringing of hands. This is a moral panic.

A moral panic is not a political reform. One can tell a true moral panic by its political achievements. There aren't any. Nothing of consequence changes. During the panic, some person or group is usually scapegoated and severely punished. But when the panic fades, no one is happier for it; no one feels any safer, more assured, or more at ease. The government suffering the panic does not become more equitable or more efficient; no injustice is actually redressed; nothing works any better or makes any more sense. No pressing crisis is fixed, settled, or improved. There has merely been a brief public orgy of hair tearing. Another, similar panic on the same topic can occur under the same circumstances at almost any time.

Moral panic is the signature political motif of the information age.

Moral panic haunts a networked world for a good reason. There is very little else that can make any difference.

Consider the Internet. It is the ne plus ultra of creative destruction, the fastest technical transformation in human history. Even its double-digit growth rates had double-digit growth rates. It falls down quickly and destroys huge fortunes overnight. The Internet is not an information superhighway. Superhighways are build on public land by governments, run by traffic laws, and thoroughly policed. The Internet was built from the bottom up, almost accidentally, without a master plan and in great haste. It connects millions of people in dozens of subcultures across scores of national borders. It is also very poorly policed. Spies lurk everywhere, and the police focus on seizing the computers of malefactors, but no one has formal jurisdiction over this vaporous realm. There are no established legal Internet traditions. If you go out to "find some Internet," or "buy some Internet," or "sue some Internet," there's nobody there. There's no Internet, Inc. It's got no president, no senate, no CEO, no stockholders. Just users and their hardware and software.

Now consider yourself as a political actor confronting the Internet. Obviously you can get your own website, promote your political ideals, encourage your friends, and harangue your enemies. You can get ECHELON to spy on transmissions, you can befriend Microsoft, and you can hope for friendly coverage on the AOL Time Warner and CNN websites. But then what? What can you do politically to affect the development of this strange nongovernmental global enterprise?

Your options turn out to be severely limited. You cannot make a firm legal argument and compel any serious attention. You lack any physical ability to control and corral the users. You can't buy them off. You can't tax them. You can't even use government funds to build the thing properly. But provoking a moral panic is a tactic of genuine promise. A technocrat who exhausts his rulebook hits the panic button. Better to light a brush fire, and perhaps provoke some useful mass stampede, than to declare oneself entirely irrelevant to the course of events.

The Internet's users are educated, well-to-do, technically sophisticated New World Order people who have vast amounts of information at their fingertips. On the face of it, they look like model citizens for the years ahead. One might naturally assume that this self-selected technocratic elite would be very sober and judicial. They should lead the way to a new global politics of precise performance and chromed, machinelike professionalism. But the exact opposite is true. What passes for Internet government and Internet politics is all fads, hype, plagues, scandals, sudden jolts, and muffled, distant hammering.

Computer activism is a mother lode of moral panic. Although there has never been much in the way of actual terror on the Internet, it exults in rumors of terror. Internet activism of all ideologies is routinely dominated by bizarre worst-case scenarios, as small groups of zealots scream imprecations and wave a wide variety of bloody shirts.

Internet pundits routinely employ fantastic, dizzying prose in even minor matters. People are used to this level of hype in computer-stock promotion, where everything seems to be standing still if it's not doubling in eighteen months. In a political context, the Internet is Panic Central.

For years on end, the political rhetoric of cyberspace was chock-full of monstrous phantoms. They are deployed like glossy *Star Wars* trading cards. The fearsome quartet of "terrorists, child pornographers, drug dealers, and the mafia" has been cited so many times that they became known, with a blasé shrug, as the "Four Horsemen of the Infocalypse." Of course, real-life terrorists are almost always drug-dealing mafia as well. Pornography and endangered children were always flung in to thoroughly muddy the issue.

The Internet's moral panics come and go at electronic speed, leaving scarcely a trace on the body politic.

For instance, let's consider the—still largely imaginary—prospect of Internet terrorists. Consider the scary idea that terrorists might use the Internet secretly, to store and forward information on how to build nuclear bombs. This is a common argument among people alarmed by the Internet's lack of security and public oversight. A terrorist nuclear bomb is a severe and dramatic threat to our well-being, so it really makes people sweat.

But that is a moral panic, not governance. Nothing happens after you make this atomic-terrorist argument. Because there is no imaginary network of international Internet censors that could stop this process from happening. Finding terrorists and confiscating their atomic-bomb plans is politically, socially, and technically unfeasible. No such reform can take place. When a gentleman like Osama bin Laden needs to smuggle his real-world plans for a terrorist atomic bomb, he can simply hide them in oil pipelines, or inside a ton of heroin. A nuclear-proliferation threat most definitely exists, but the moral panic merely conjures up digital bogies—it cannot come to grips with the problem. If you're serious about frustrating atomic terrorists, it's impractical to police the flow of data. Total censorship was beyond the almighty KGB, way back in the era of fax machines. It would be much easier and simpler to find and secure all the uranium in the world. Real-world security experts have attempted this in, for instance, Iraq. Even then, it's never as simple as it sounds.

Let's take the second specter of moral panic: child pornographers. These people have remarkably disgusting interests, but this doesn't make them a threat to global security. They're a threat to adult equanimity, and occasionally to children, but not to civilization. Pornography doesn't topple empires or cause nations to collapse. But it's a classic hot-button issue. It's always there, always provocative, and never resolved.

There's only one urge as powerful as the urge to sneak a look at pornography. That's the urge to try to see to it that other people can't see any. These are both appetites that grow with the feeding. As long as we have a libido, we humans will never get over a prurient interest in sex, and we will never get over the panicky feeling that somebody some-where is having some indecent fun denied to us. Bluenoses will never stop pursuing pornographers, not even if every man, woman, and child on the planet is shrouded head to foot in a *burqah*.

Drug dealers use the Internet, a fact people find horrific. Drug dealers sell drugs because people buy drugs, not because people have e-mail. Areas genuinely disordered by drug traffic, such as the Mexican state of Sinaloa, are not major Internet centers. When it comes to media, Sinaloa much prefers its native *narcocorrida* folksingers.

Organized crime is a severe, real-world threat. But bat-tling it on the Internet is at best a sideshow. Organized crime never gets thoroughly well organized until it becomes a par-allel state and is running governments. At that point, you can no longer stigmatize it in order to obtain some passing advantage in matters of Internet policy. You will find your-self confronting deeply ingrained problems of systemic political corruption: protection rackets, bribery, tax evasion on a massive scale, and oceans of black money. Once that has happened—and it does happen—you can forget the web-surfing and the e-mail. They're irrelevant. The mafia can have all the computers it wants when, as in the cases of Arkan and Catli, it becomes the secret police.

So much for the specters of the infocalypse. But security mavens are not the only ones who thrive on moral panic. Libertarians make similarly exaggerated claims. Computer files could empower mass enslavement of the Jeffersonian yeoman by evil corporations. Yes, they could; but if corporations are inherently evil, then you're fighting on the wrong battleground. In the hands of fantastically malignant capitalists, every form of hardware is a threat to liberty; you won't help matters much by denying them spreadsheets. Go get their cars and telephones. Or their money, if you dare.

"Infowar" is supposed to kill people with the press of a button. So do truck bombs and cruise missiles, and unlike information warfare, they're not nine-tenths imaginary.

While its politics thrive on moral panic, the Internet continues to spread. It was designed in an almighty hurry and built without a garbage can or an off switch. Everybody talks about getting the population onto the Internet; there is no research at all about getting the population off the Internet. The fiber optics hum around the clock, and the lights are always on. The Internet has even begun to rot; dot-coms have vanished utterly, with websites left empty. The World Wide Web is crammed with derelict, abandoned files; the Internet is showing its age even before it comes of age. It's not as important as the discovery of fire, but it's what's big and new. It will supply the next century with a unique and native character.

Though there has never been an Internet Chernobyl or an Internet Bhopal, accidents happen. The unprecedented structure of contemporary media enables some activities that were formerly unfeasible and impractical.

"Surveillance" is a pejorative term, but surveillance is media. Surveillance and media are the same thing. Better media, better surveillance. Surveillance does not exist just for the delight of Big Brother. "Big Broker" is a likelier candidate nowadays, and the Internet excels at making the invisible visible. Video cams have become tiny; they go unnoticed and unresented. Facial recognition by software and cameras has become a growth industry. Our telephone bills can finger our friends and relations for the delight of investigators. Credit-card transactions paint a detailed portrait of our location and economic interests. Our online purchases are cataloged, sorted, and data-mined. Medical records detail our pain and weakness. Global-positioning locators watch our cars. Our portable phones can track and record our movements from cell to cell. Scary genetic scans suddenly abound in police investigations, so we find ourselves leaving unsuspected trails of our own shed DNA. We can be smelled back to our very lairs by anyone with a sequencer and a cotton swab.

Greater technical power gives a society a competitive advantage, but it does not make that society more stable. By multiplying our options, technical power magnifies our hidden eccentricities. Naked power looks extremely naked when it is subjected to surveillance. Things traditionally hidden can blow out like a tire under pressure.

Imagine being publicly transformed into a pervert, quite against one's own will and intention, as part and parcel of a vast, lubricious, septic media orgy, on millions of screens around the world. It's a heroic agony that would require iron will to endure. Yet it happened.

President Clinton's impeachment was a definitive moral panic. This was an out-of-control media circus of the first order. Because of its novelty and its apparent unlikelihood, the Clinton impeachment scandal is probably the best harbinger we have for the native character of twenty-first-century politics.

Nobody wants to go through that again, but nobody could have predicted it and ducked it in the first place. A massive, thoroughly crazy sex scandal could happen again, at any moment, to almost anyone in power in any nation. There are no effective boundaries or safeguards to prevent it. Nobody has invented such safeguards, and it's hard to imagine what they might look like or how they might be put into place. Meanwhile, the arsenals of surveillant technology that enabled the scandal are getting stronger and more widespread every day. The Condit scandal was a smaller replay of that phenomenon, visibly struggling to fill every minute of dead airtime. There is every reason to believe that this process will happen repeatedly.

We might call this "network toxicity." Not shiny, antiseptic, Big Brotherish, and high-tech, but incredible, ludicrous, bizarre. Office gossip, a dalliance, recorded phone calls, websites, lawyers, and cable TV leading to a constitutional coup.

It never seemed efficient or machinelike; it grew like bread mold, with organic profusion.

In retrospect—these things are always easy in retrospect—the disaster showed warning signs. In 1987, when Judge Robert Bork was nominated for the U.S. Supreme Court, enterprising enemies discovered and revealed the records of his videotape rentals. They were, of course, looking for pornography, in the hope of springing a moral panic. Had they found any, Bork's nomination would have become an instant debacle, probably rivaling the bizarre sexual circus of Judge Clarence Thomas some years later. Judge Bork, to the disappointment of his persecutors, had no fondness for video pornography. However, Congress was so horrified at this attempted porn smear that it successfully passed the Video Privacy Protection Act of 1988.

This legislation made American "videotape service providers" liable for civil actions if they revealed their rental records. Here government was responding vigorously to a clear and present danger to political stability. But this act was a stopgap at best. In an age of DVDs and streaming media, what the heck is a videotape service provider?

In an age of freely available, booming Internet pornography, the sticky fluidity of sexual data is not to be restrained by mere congresspeople. By 1998, a mere decade later, the network revolution had advanced so marvelously that the Congress itself deliberately piped highly salacious details about oral sex over the Internet. Same impulses. Same motives. Far better technology.

The reaction surpassed all imagination. The American government was stricken as if by voodoo. The major figures in the drama, portly men of dignity and gravitas who should have had the stylish solemnity of Shakespeare's judge, were transformed into a gang of smutty comedians. Even television talk-show pundits, trained to make almost anything seem credible, were taken aback. Seemingly enchanted, Washington plunged week by week, month by month, into stygian depths of utter prurience. Historians will marvel. Not because it won't be happening to their own society, though. Probably because the public reactions of the 1990s will seem so quaint and naïve.

Sex is an absolute given in human affairs. Without it, there can be no humans and, of course, no affairs. Quite naturally, it turned out that the president's foremost political enemy, the Republican speaker of the House, was carrying out a far more reckless and indiscreet extramarital affair of his own, even as his subordinates plotted the president's downfall. After Speaker Gingrich fell from power, the following speaker of the House was also tripped up by a sex scandal, surviving politically for only a few days. Politicians somewhere are having illicit sex even as I type these words. Politicians are affable, charismatic, and persuasive by nature. They will never give up seduction, any more than athletes can stop flexing their biceps.

It was the unexpected, unstoppable, mediated propulsion of private acts onto the public stage that warped politics and damaged the status quo. It's hard to imagine how we can step back from such a situation. We can't disinvent the hardware. Asking people to forsake their network connections is like asking them to become illiterate for the public good. And they'll certainly never give up having sex.

Once data has been blown to thousands of users across dozens of national borders, it cannot be successfully recalled or expunged by anyone, anywhere. The previous game of gentlemanly discretion and double standards cannot return, because the old system of gated, hierarchical information flow is gone. Despite horrific losses in dot-com stocks, the old oak-paneled Wall Street cannot return, either, for millions of people worldwide can trade stocks anywhere, around the clock.

Perhaps the best hope is that this particular form of moral panic becomes passé from overuse, even though human prurience is as sure as the rising sun. People do seem very unlikely to shrug and look elsewhere when important political leaders have their pants down. Really, how could they? Sex scandal is a byword for riveted media attention. It's the one political act almost guaranteed to draw a crowd.

Will people somehow learn to watch their leaders rutting, with a sophisticated twenty-first-century chuckle? Can that data be fought with more data? Can it be spun, denied, discredited, distorted, or ignored? Perhaps the best response to being caught having sex with your intern is to claim you've had sex with ten thousand interns. Perhaps you could court voter popularity by cheerfully offering to have sex with all of them. Are such things even possible? We'll see.

We live in a world of disruptive surveillance, but it's not *1984*. In George Orwell's science-fictional concept of a wickedly networked society, the sinister Party watches the people on every street corner, and doublethink is imposed from on high upon the mind-deadened proles. Technical reality has moved on from that scenario. A network is not hierarchical; it's not disciplined, pyramidal, gated, and machinelike. It is distributed and swarmlike. That swarm-like, scattered structure enabled the Internet to cross national borders. Networks visibly sponsored by governments (such as Minitel) did not do well. Those controlled by business interests (such as CompuServe and Prodigy) tended to stultify.

In today's configuration, networks are not tools of harsh public discipline. They don't bring us rigid Orwellian order; they are far more likely to create repeated, seriocomic, sur-real explosions. Farcical bursts of utter weirdness. Fads, hype, plague, scandals, sudden jolts. Red tides and dripping slime.

Digital networks have a technical stamp of American society and of American social values, but network toxicity is a global ailment. To be a British royal has become one of the most hazardous and humiliating jobs in the world. The

"War of Compromats" among Russia's media oligarchs was debilitating and extreme, an amazing orgy of charge and countercharge by moguls struggling to ruin one another's public reputations, even as they bought and ran the Russian media. In their televisions, magazines, and newspapers, they libeled one another so ruthlessly and successfully that two of the miscreants finally fled offshore to escape arrest. Since these media moguls were also the prime courtiers of the Yeltsin family, the Russian experience probably cost that country a president. It likely had plenty to do with the fact that the next Russian president was a career KGB spy.

American politicians sometimes refer to media toxicity as the "politics of personal destruction." But this is not just destructive malice at work; the people in politics haven't suddenly become more evil. On the contrary, they've never been so dull, modest, and practical. But they do have some incredible hardware.

This personal-destruction problem is better understood as a politics of privacy rupture. America now combines hugely overfinanced campaigns with excellent digital databases. That's a very toxic mix. Opposition-research experts haunt every professional campaign, researching and compiling any event in a candidate's life that might be spun as a damaging flaw. The outcome of this rampant digital investigation is not cleaner politics or greater public accountability. Far from it. A new broom sweeps clean, but if every political candidate is attacked for every peccadillo in his or her life span, nobody's broom is new and nothing can ever be clean. It's like having one's candidates ritually beaten to death with brooms.

People in America have become savvy about networks and have assimilated and domesticated them to a remarkable degree. It's hard to recall now how utopian and abstract computer networks once seemed. But technology marches, while the law crawls. As the girders and curtain walls of the network economy fly up (and so often collapse), there's a bland expectation that political developments will somehow work out for the best. Why should they? Have they ever?

Life's not all hype and vaporware in Netland. People of intelligence and goodwill have devoted serious attention to this problem. A stellar example is Lawrence Lessig, a Harvard and Stanford law professor and author of the book *Code and Other Laws of Cyberspace.* In this work, Lessig chose to think outside the legal box, showing that his footwork was just as fast as that of any computer mogul. In classic Silicon Valley entrepreneurial fashion, he suggested a new, network-literate model for future government, boldly declaring that software should be seen henceforth as the "law" of computers. "Code is law."

Lessig argued that merely legal attempts to regulate computers and networks are doomed. "Courts are disabled, legislatures pathetic, and code untouchable." The Internet moves too fast, its architecture is a hodgepodge, and national jurisdictions are always tripping over its international scope.

So, surmises Lessig, what the new century really needs from its native politics is not traditional land-based government but a newer, subtler, techwise form of network-friendly "governance." This digital governance would not be nailed down to any mere real estate; instead, it will dominate the code-writing process in the public interest. In this virtualized republic, a politician would become a kind of constitutional interface designer.

This is a strain of political thinking quite new in the world. It's been thought through soberly and in earnest. If it sounds wacky, that's because it's coming to grips with a wacky situation, a world order where information seems like everything to everybody: it's commerce, media, politics, science, art, education, military power, a good, a service, a dessert topping, a floor wax, and weirdest of all, an amazing gush of rampant, lubricious pornography that ran straight up to the White House, Senate, and Supreme Court.

When the going gets weird, the weird turn pro. Lessig's core idea is to crossbreed Washington's oversight with Silicon Valley's rapid development. My good "code" outdoes your wicked "code" because the U.S. government smiles upon it in the early days of the lab. So my good and healthy code gets a critical early edge in the technological adoption process. Once these wise, approved standards have been firmly set in place, "technological lock-in" occurs. In a technological lock-in, a complex system is already established everywhere in sight, sucking up all the creative energy and all the market oxygen. It's very much like the global reign of Microsoft Windows—except this time it's run by sensible public servants, rather than by some weird rich guy named Bill.

Of course, there's an inherent danger here—because once this lock-in is engineered into place, all the lawyers in America will find themselves huffing and puffing at an immovable house of technical bricks. If Code Is Law, then code is that which rules, while traditional law is reduced to a polite pretense, a feeble gloss on raw technological power. Where does power sit once the lawyers are ornamental potted plants?

The commercial agents of a lock-in are multinational. They have a hammerlock on titanic sums of money. They can afford better lawyers than a national government can. Or they can just buy themselves a nicer government.

Let's suppose that government wises up, gets the picture, and deftly moves these locks into place. Once that happens, other alternatives vanish, quietly and discreetly. This clearly offers great benefits. Certain frowned-upon actions in digital networks (such as, say, anonymity, spam, porn, identity theft, market manipulation, viruses, digital espionage) need no longer be outlawed. They've been designed out of existence, rendered impossible.

To give him its due, Lessig has come up with a practical, hands-on, and very contemporary vision. This is a step well beyond empty panic-mongering. This is deep creative thinking under novel circumstances, and just the sort of thing one likes to see from a scholar. There's daylight here, there's debating room.

But for traditional fans of justice, liberty, and democracy, there's not much comfort. Government supposedly derives its just power from the consent of the governed. A Silicon Valley start-up is nothing like a democracy. There's no voting. There are no checks and balances. There's no Bill of Rights.

The outer reaches of software and network development are inherently arcane and elitist. They're not democratic because they can't be. If every voter was fully informed about every new technical standard, it couldn't be new or even technical. Besides: aren't these supposed to be absolute issues, self-evident truths, the moral bedrock upon which a free people chooses to build its society? Human rights are supposed to be rights, not browser plug-ins. It's "liberty and justice for all," not liberty and justice as a set of configurable check-boxes.

What is government to do about hydra-headed, rapidly moving, multinational technologies? They are a powerful source of competitive strength among nations, and a menace to stability within them. Unless statesmen can bustle in with blueprints and design those innovations themselves, they'll get hammered again and again by forces beyond their reach and understanding. There will be moral panic after moral panic, as things considered unthinkable or beyond the pale are presented to a stunned and sweaty global public on silver technoplatters.

There's a grave pacing problem here. Microsoft boasts as a matter of course that every product it makes will be obsolete in four years. Software dies fast and it gets no burial. How much time and energy can courts and legislatures waste on that dizzy process? By the time you legally define what a "browser" is, the "browser war" is all over. By the time you pass capital flight laws, your economy has collapsed. By the time you break up your national phone monopoly, the Finns and Japanese have invented entirely new phones. By the time you protect Judge Bork's videotape rental records, nobody uses Betamax.

Thorniest of all: the real-estate problem. National legislators are elected by states and precincts—the good old-fashioned dirt world. The citizens who elect them are born of the homeland, children of the national soil. Politicians are archaically based in a merely territorial sovereignty. The parochial goings-on within their courts and legislatures have less and less to do with the daily lives of their constituents, who are wearing Taiwanese shoes and drinking Brazilian coffee.

National politicians are in the same basket as contemporary labor unions, who can't control the global flight of capital that exports their jobs. They don't have the carrot of controlling national prosperity; they don't have the stick of military drafts and massive land wars. In twenty-first-century

Europe after the dawn of the euro, states don't even have national currencies. It's little wonder that their decisions have less relevance and therefore command less respect.

In an age of networks, it's cheaper to move technology and money than it is to move people. People feel a patriotic qualm about "hollowing out" their own country, moving their technology and money offshore, or going abroad to work when things look bad in the homeland. But that process is always powerful, for everyone is always pleased to hollow out all those other countries.

Therefore, the money moves away whenever some uppity national government politically restricts some activity in the name of its people. If you're having a war, you want to have an American-style New World Order war, the kind that drags every major commercial actor into the ranks of a global coalition. It's not that you require the firepower of distant Australia and Japan, but you do want them inveigled into the hostilities, so that capital will have no safe place to run. Otherwise, the money flits as if by magic to someplace more open and understanding, in some place that's lighter, faster, more connected, more megabits per capita. For outsiders— and most of us are always outsiders to most of the world—a nation is just another brand name. Buy China, hold Finland, sell Indonesia.

Mind you, people, we citizens, can still be arrested and jailed by governments (nowhere with more gusto than in the world's leading Internet pioneer, the United States). The money goes, but you don't get to follow it. If you could, then the money would have no reason to run; it would have to sit where it was with a sense of grim resignation. The money is running from you—from you and your decisions that affect its interests. Citizens of nation states are meticulously labeled with passports, scanned with magnetometers and sniffing dogs, and personally searched down to their shoe soles. But money goes almost anywhere—while data moves even faster and more randomly than money does.

Can national governments think through this, recognize the new playing field, and get ahead of the game?

Maybe. There is good cause to doubt it. Consider the miserable doom of the U.S. government-designed-and-approved Clipper Chip. The Clipper Chip was created to make communications secure from attack by, well, terrorists, pornographers, drug dealers, and the mafia. It was designed to scramble phone calls and computer communications, so that no one could listen in—unless, that is, they were federally approved eavesdroppers from the United States government. American spies and cops had no problem listening in on the Clipper Chip. In fact, the Clipper Chip was designed on purpose so that Americans could secretly listen in.

Why anyone in France, China, or Iran would ever want to buy an American spy device was never explained. Designed by agents from the National Security Agency, the Clipper Chip was of course about national-security interests. No foreign customer anywhere was dumb enough to buy it, and the U.S. government lacked the nerve to install it by decree in American phones and computers. So the Clipper Chip was rejected worldwide, instantly, with virulent contempt. No other, better solution was allowed, though. So communications remain highly insecure to this day, to everyone's general hazard. This is a great example of how to bungle a technological lock-in.

There is one last big dilemma in earnestly pursuing democratic issues of public policy in computers, freedom, and privacy. It's the spook problem.

The networks have never been democracies. As a consequence, they have always been haunted by spies. Cryptography and electronic surveillance were there at the very cribside when computers were first born. Digital computers were invented in order to crack Nazi clockwork code machines. In a strange moment of Lessig-like rationality, once World War II was over, Winston Churchill brusquely ordered that Alan Turing's code-cracking computers be hammered into chunks and secretly dumped in the North Sea. Politicians knew how to command the field back in those days.

But the temptations of covert technical power were too strong for democracy to resist. With Harry Truman's secret founding of the National Security Agency in 1952, a vast surveillance apparatus was created for the cold-war epoch. Those networks still exist, routinely soaking up radio and phone transmissions from around the world and, of course, analyzing Internet traffic. Recently their powers have been drastically expanded in the service of an undeclared war on terrorism. But these intelligence agencies are beyond democratic argument, beyond scholarly analysis and public debate. Not merely technically obscure, they are deliberately concealed for potent reasons of national security.

National governments have a very long tradition of secretly tracking and listening to the signal traffic from other nations. Like war, espionage is politics by other means. Also like war, it is the symptom of political failure.

National governments that attempt to assert global law and order are playing beat cop and Peeping Tom at the same time.

No nation is entirely innocent here. Electronic spy networks, "technical means of verification," are lavishly funded. There's great advantage to be gained from their eavesdropping, though mere citizens and voters rarely get to hear of it. Almost every embassy has microwave aerials; nobody ever shuts them down. Even the smaller actors, who can't afford big-budget spy satellites, beg their data from allies. No one's hands are clean; it's a jungle out there, and behaving as if it's a jungle is what keeps it a jungle.

It's rather surprising that the world's best global network monitors, the National Security Agency and its very skilled British ally, can't reinvent their role to suit new global realities. They are the biggest and stealthiest predators lurking in the wilderness of international communications, but thanks to their huge eyes and ears, at least they do know what's going on. If their work was open, public, and legitimate, they would become a great stabilizing asset.

The traffic of a global information infrastructure could be monitored wisely, but it would require serious political action and informed consent by the major parties concerned. Building workable, trustable networks, even in good political faith, is very hard, and certainly can't be done in pitch darkness from a secret fortress in Fort Meade, Maryland, when the civilians aren't allowed to look.

An ideal NSA would become those Lessig-like architects. It would be planning and building and carefully watching that healthy, life-affirming public network in the full light of day. But it's just not up for this difficult work of global citizenship. Because of this grave systemic failure, it is constrained to live and work like a hacker, surreptitiously cracking and breaking, installing its trapdoors and back alleys.

Occasionally someone catches on, as the European parliament did when it investigated the ECHELON monitoring system, or as the German military did when it rejected Microsoft security products as riddled with NSA trapdoors. But most of the time when people trip over the peculiar artifacts of this surveillance demimonde, they simply dust themselves off and go on as if nothing happened.

This situation is very destabilizing. It breeds a well-founded cynicism. It's a New World Disorder situation on a global scale. Invisible antiterror groups deprived of democratic oversight are especially menacing. They are the proven breeding grounds of Arkans and Catlis, Teflon-coated agents of corruption and mayhem who subvert their own governments.

Among many other ugly consequences, arrogant electronic espionage means that democratic legal scholars are robbed of their legitimacy. When Lawrence Lessig and his colleagues appeal to a higher social purpose and ask the Internet's technicians to wisely plan for political consequences in the public interest, they know that, as lawyers and servants of justice, they can't deliver the goods. The U.S. government and its federal attorneys are not honest brokers in global networks. When it comes to telecommunications, the NSA has a permanent veto over the Justice Department, thanks to a time-honored eavesdropping that has been unconstitutional and illegal from the get-go.

An honest and informed debate about a networked republic or a networked planet is impossible under these circumstances. This stymies all other political advances in that arena. The result has been a slow shadow war of computer geek and computer spook, where spies and cipher enthusiasts try to trip one another up with technical faits accompli. It's another toxic network struggle, dominated by leaks and scandals—except in the special case of the eldritch NSA, the moral panic comes in a billowing fog of dry ice straight out of *The X-Files*.

Deceit is the sorriest element here. It undermines every

other effort. Espionage is a form of power immune to democratic legislation. It is also a tremendous wellspring of network toxicity, because it is secret, anonymous, and surreptitious; it's a fabulous, unending source of privacy invasion, political subversion, damaging leaks, dirty tricks, and every kind of black propaganda.

It would take a mighty effort of reform for those in power to lay this wondrous temptation aside, forsake all this preciously hoarded secret knowledge, put the public issues on the public table, and decide to settle for mere, straightforward, accountable government in a fully networked context. That would mean global glasnost and perestroika, and heaven only knows if we'd survive it. Yet until that's done, there can never be civil legality in the networks. They will always have vast damp patches of shadowy corruption. They will never be able to move into the full light of day.

Adding a global war on terror to the Net is sure to fertilize its worst eccentricities. Computer-science notions about infowar, once nine-tenths theoretical, will look far more hazardous within a paramilitarized Net. If cyberspace becomes the stomping ground of armed secret agents, that is sure to be corrupting and unlikely to be healthy for the innocent user. Many of the Net's pervasive vices—software piracy, viruses, online casinos, mujahideen websites—will look less like colorful anomalies and much more like excellent motives for some Catli-style liquidations.

When will the politics of a networked society achieve

stability? Is justice possible here? Even if we finesse the direct and crazy violence of the erupting disorder, it may well be that we're doomed to a long-term political environment that combines stunning boredom with spasms of dizzying nutti-ness. A world where everyone in power pretends to be in-sanely normal, technically adept, and thoroughly in charge, until some random bit of media grit—a lame joke in front of the microphones, a mispronunciation, a kiss, a drink, the wrong color tie—bloats onto the public stage, all damp and spiny, like a poisoned puffer fish.

Why should people settle for that situation? Even if net-works are inherently organic in their behavior, this shouldn't make one lose hope of positive political action. A garden is organic, but that doesn't mean that a gardener has to love crabgrass. A language is organic, but that doesn't mean you have to lie.

What does the future of politics hold? Suppose that, as a society, we move beyond technocracy, nostalgic activism, and the bizarre. This means imagining a new form of politi-cal activism that is not nostalgic but futuristic. It's hard to outguess politics—every political activist always wants to outguess the system, and there are millions of them—but we can speculate.

If the New World Order fails and collapses, then any-thing is possible. If technocracy cannot deliver on its solemn promises of economic growth, if its torrent of consumer goods dries up, and if the screens of its bewitching entertainment go dark, then truly ferocious political energies might be unleashed.

There's no reason to assume that the twenty-first century is inherently more sensible or less bloodthirsty than the nineteenth or twentieth, and for a global world based on mercantile prosperity, a Great Depression would be an ultimate disenchantment. The bizarre would rise toward power with torches in hand.

Technocracy's dominance is firmly based on a general conviction that political activism isn't likely to get you anything worth having. Wherever technocracy thrives, voter turnouts dwindle as a matter of course.

But we're not assuming a breakdown of civilization. We're trying to look for a workable, real-life yet novel form of politics that might plausibly grow from today's trends and circumstances. What might that mean? How would it look and feel?

Here are a round dozen suggested signs for a new twenty-first-century political movement:

First, this movement would need a genuinely new ideology. A novel, galvanizing Big Idea, something sloganizable, along the lines of "Liberté Égalité Fraternité," "One Man One Vote," "No Globalization Without Representation"— that sort of thing.

This ideology need not look traditionally political. It

might seem rather goofy and eccentric at first, like, say, femi-
nism. It might take us quite a while to realize that the pro-
genitors of this movement were not bizarre, that instead,
they had thought the matter through and were serious about
their issues. As time passed, they would find themselves winning some important arguments and attracting serious-
minded adherents.

This political movement is likely to be proglobal and
multilateralist. It's unlikely to base itself within a single
nation-state, since national governments are severely bot-
tled up and appeals to local patriotism are self-limiting.

It will need some physical strongholds and some model
polities. National states don't seem particularly promising
here, at least not at first. A likelier candidate is large cities.
The governments of cities can be captured by small up-
start groups of enthusiasts, and the best such candidates
would likely be multiethnic cities, heavily involved in global
trade and populated by diasporas. Brussels might be quite
good. Singapore. Perhaps New York City, Amsterdam, or
Hong Kong.

The key to success is that these cities must put that new
political doctrine into practice and then find that people are
flocking there by preference. Their proof of concept would
be new governmental policies that bring greater prosperity
and an improved quality of life. If they are winning, then the
strongholds of this movement will be widely perceived as
more civilized, more sophisticated, more entertaining than

the planet's backward areas. The globe will vote with its feet, in their favor.

This movement would very likely be led by quite rich, sophisticated people. If there's anything that the recent unpleasantness with Osama bin Laden has demonstrated, it's that you no longer have to be poor to be radical. On the contrary, bizarre ultrarich characters such as George Soros, Silvio Berlusconi, Steve Forbes, Ross Perot, and, yes, Osama bin Laden have profoundly disturbed the political landscape. Soros with his Open Society initiative has the closest to a genuinely innovative political ideology—but all movements nurtured by single founders will suffer from Napoleonism.

This movement would have to be different. It would involve the very rich but would finesse that wealth problem— perhaps by forming a Grand Diaspora Alliance between the jet set and economic refugees.

The very wealthy care little for nation-states; the dispossessed fear and resent them. The poor badly need capital, while the rich badly need voters. This looks like a potential global top-to-bottom alliance. It would probably take an alliance that unorthodox to crack the power structures of the current status quo.

If capital moves across the globe and is swiftly followed by a huge horde of rootless people who are somehow enfranchised by that money, it could mean a new coalition of genuinely globalizing social forces. This could overcome the traditional political problem posed by real estate. This

prospect is a little far-fetched but not entirely without precedent; it worked very well for Hong Kong. It even worked for
gold-rush California.

The twenty-first-century world is a crowded one—one
might think there's no room left in it for newly founded
societies. But the disorder offers plenty of elbow room.
There are many handsome places that people have abandoned in despair. Beirut was quite a lovely place once; Sara-
jevo had a great deal to recommend it. If a group of daring
global investors offered to rebuild Kabul wholesale, they'd
likely be met with open arms.

Another potential signifier: the primary supporters of
this movement are too young to realize that politics aren't
supposed to work. Too naïve to be cynical about activism,
they are firmly convinced that they have invented politics.
The "politics" their grandparents talk about dismissively was
just done all wrong; it had the wrong attitude. When these
new players hear that it's the Summer of Love in distant
Uzbekistan, they don't tell one another how unlikely this
seems; they just pack up and go.

This movement may find its first power bases outside
nations—in cities, within NGOs, and in global enterprises
in and out of the public sector—almost anyplace not poisoned by traditional political exhaustion. However, the likeliest sign of their success will be their passionate interest in
voting. Voter turnouts are so low in technocracies that any
genuine broad-based enthusiasm ought to dominate the

polls with ease. A new political movement that was bound for glory would not be queasy about voting. They'd go door-to-door. They'd hold big public rallies. They'd assemble precinct captains and political machines. They'd have the period equivalent of torchlight parades and podium-thumping beer busts.

So let's sum up this whole speculation. It's a new movement that recruits from a diaspora, has offshore strongholds as bases, unites the very rich and the very poor, has globalizing ambitions, is passionate, uses lots of naïve young zealots, looks parapolitical and rather NGO-like . . . Basically, this organization looks like al-Qaeda. That's no accident. The al-Qaeda terrorist network is a global political movement attempting to succeed under native twenty-first-century circumstances. Unfortunately, it's a crippled botch and was very badly designed. It's run by murderous autocrats, it has no formal way to derive the consent of the governed, and although people do flock to join it, they are almost all bitter young men with guns.

An al-Qaeda that could turn the homelands of disenfranchised Muslims into stable, settled, prosperous Casbahs and golden-age Baghdads would truly be a force to reckon with. It could pretty well write its own ticket in the Muslim world, where people have never quite gotten it about nation-states, anyway. The world has a billion Muslims. The Muslim diaspora does rather well for itself—when Muslims are allowed to live under circumstances not riddled with corruption and

raw intimidation. Salman Rushdie, for instance, is quite pros-
perous and sophisticated, a justly influential cultural voice.
He seems at least as much at his ease in terror-stricken New
York City as he is in his hometown, terror-stricken Mumbai.

Much the same can be said for the Chinese diaspora, and
the Indian diaspora, and even the Russian diaspora. These
are all people of considerable ability whose dysfunctional
national politics are visibly holding them back. They have
great commonalities, similar advantages, similar resentments.
Nonresidents, with their flow of remittances and hunger for
home news, are also becoming persistent and deeply moti-
vated Internet users. Perhaps someday these rootless, globalized groups will realize that, since they already have the
brains and the money, they might as well invent a new power
structure to suit their joint interests.

It could even be argued that the U.S.A. is the first great
diaspora society. It is a vast landmass where ancient hatreds
and stultifying traditions were deliberately shorn off in pursuit of a headlong return on investment. This polity now
dominates the world. But the U.S.A. also has one of the oldest and most continuous governmental traditions among
modern great powers and has jealously settled into its role as
a continental nation-state. Its population is stable, it is top-
heavy with the old, and the Internet is a hobby for American
culture, which already owns Hollywood and CNN. The
U.S.A. has too much to lose to justify great outbursts of
inventive political radicalism. But the globe is a big place.

Demographics are shifting. Education and access to data are spreading every day. The clock will not stop ticking. Though people are sometimes trounced by terror or moral panic, they can also be inspired by symbols of hope.

There are solid, sensible reasons to expect good things of future politics. Life may seem dark at times, especially when darkness is cruelly thrust upon us through grand mediated gestures. But people have lived worse lives than ours in the past, and undergone far greater trials, and even committed far greater evils, and they have survived. They have even prevailed.

If there's a single grand political symbol at the turn of the century, a splendid and shining exemplar of justice renewed, progress restored, and an axis of evil brought to a conclusive end, it's not in America. It's in Germany.

The very name "Reichstag" is enough to send a chill down the spine of any veteran of the twentieth century. And for excellent reason.

The luckless Reichstag was designed in 1882 by a German architect, Paul Wallot, in an odd mélange of Italian high renaissance, Gothic, and baroque stylings. As a kind of industrial afterthought, it also featured some high-tech German steel and a big dome. It took ten long years to build, because various ministries and committees kept horning in on the action, constantly revising the plans. Once it was completed, Kaiser Wilhelm insulted Paul Wallot and deprived him of his imperial medals.

Wallot then spent a further ten years decorating the structure with paintings, sculptures, and statues, many of which were loathed by the Reichstag's ten major political factions and thirty-nine nutty cult groups.

In World War I the German military took over the building and used it as a center for censorship and propaganda. After the German defeat, violent left-wing Spartacists seized it in 1919 and looted it thoroughly. The next regime turned the Reichstag into a fortress. But much, much worse was to come.

If that New World Order goal we were discussing earlier— a stable capitalist nation-state, with transparent accounting, low inflation, price stability, et cetera—seems a little colorless and vapid, then it's a grand idea to have a good look at the polar opposite of that kind of country. That would be Weimar Germany. It was a crooked, incompent, violent state with catastrophic inflation and crazed politics.

Weimar ineptitude made extremism fatally plausible. The Weimar parliament was thrown out of the Reichstag by Chancellor Adolf Hitler on his second day of office. The parliament was abolished, and in 1933 the Reichstag conveniently caught fire. The Nazi regime, after cleaning up the blasted wreck a little, used the structure for anti-Semitic agitation and Goebbels movie festivals.

In World War II the Reichstag became a Luftwaffe base and a Wehrmacht bunker. Allied air raids damaged it severely. Then came the turn of the Red Army, who shelled it with the utmost severity.

The blackened Reichstag rotted until 1954, when the locals in divided Berlin took the trouble to blow up the scary wreck of its glass-and-steel dome. In 1961 the Berlin Wall was built almost at the Reichstag's doorstep. Lackadaisical repair efforts got the Reichstag more or less complete again by 1972, when it became a historical institute. It had plenty of history to teach, but very little of it was comforting.

The Reichstag is a building to which nothing has ever come easy. It is a cursed and snake-bitten structure, a place where a nation's best energies have often been tragically misspent.

When the cold war ended, Germany reunited, and the decision was made to return the capital to Berlin. It took some creditable nerve on the part of the Germans to hire a British architect to rebuild their Reichstag. It must have taken almost as much nerve on Norman Foster's part to accept the job.

Lord Foster of Thames Bank is certainly one of the planet's greatest living architects. He has awesome technical mastery. He also has a first-class office of four hundred designers and architects, plus a favorite engineering firm. The Foster atelier has created and completed some of the world's largest and most ambitious projects, from the Hong Kong airport to the tallest skyscraper in Europe.

But even a wizard like Foster had no magic wand for a historical mess as bad as the Reichstag. It took him two years to design the new Reichstag and four years to rebuild it. True to form, it was a bureaucratic slog all the way. There were setbacks, there were compromises, there was peevish and pointless obstruction. There was, in a word, real-world politics.

Everybody had a beef. They didn't like his modernized version of the German eagle. They were at sixes and sevens about the new glass dome. The construction crews even had to wait with bated breath in 1995 so that Christo could wrap the building. But in 1999 the renewed Reichstag formally opened. As Premier Gerhard Schroeder said, "I want this glass dome to become a symbol for the openness and transparency of our democratic politics." The building is physically secure, but as Foster says, "it is a building without secrets."

The Reichstag has been rid of its grim associations and symbolically cleansed. This is not a mere pastiche of paint and face-lifts. The structure has been profoundly rethought. The end of the twentieth century has finally given us the definitive Reichstag, the Reichstag that surpasses all others.

The new interiors are carefully joined with past structures: shiny new doors are set precisely into cannon-scarred rock. Slender frameworks of steel and glass, which saturate the new Reichstag with open daylight, zip along under the old stonework. There are many burns and shrapnel marks from the old horrors, respectfully retained. Even triumphant Soviet graffiti has been preserved here and there. This is the old, deliberately allowed a role in the future.

The Reichstag is now the greenest capitol building in the world. Lit by daylight and ventilated with fresh air, the whole shebang runs on solar panels and vegetable oil. It even supplies clean power to nearby buildings.

Its grand new themes are lightness, transparency, permeability, and public access. The new domed roof is open and accessible. The glass dome contains a giant reflective lantern that carries sunlight down into the debating chamber. The German public strolls on catwalks around the dome's three thousand square meters of glass, far over the heads of their public servants. This amazing dome, this fantastic advanced toy, all moving chrome and Teutonic precision, has been grafted seamlessly right into the stony hulk. And when you sit inside it, gazing down on those steadily complaining politicians, you can feel the whole tensile structure very subtly vibrating. To see today's Reichstag is to recognize Germany. It is truly its capitol, and it is the Germans to a T. The Reichstag has finally lived up to its slogan: "To the German People."

Eight thousand people every day pace along these spiraling, panoramic viewing decks that show the new and united Berlin: the Brandenburg Gate, the Tiergarten, the Potsdamer Platz, all those sinister places frozen through decades of cold war, now at the center of Europe.

It takes a mighty effort of magnificent focused symbolism to enlighten one of the darkest relics of the twentieth century. But it's been done. To sit inside the shining dome of the new Reichstag, watching the sun wheeling and the city thrumming, is to experience a remarkable feeling: the antidote to moral panic. It is a very strong sense of functional serenity, a mature, judicial willingness to confront error and aspire to justice. Rounded, severe, formal, and wise, the new Reichstag is a building fit for Shakespeare's judge. It is a wonderful place. It feels like a headquarters of a new and better civilization.

Like the century it's built to serve, the new Reichstag is a little strange-looking and more than a tad eclectic. But it's a throughly modern political structure that is graceful, sober, well designed, impeccably organized, and yes, it's just plain beautiful.

Better yet, the twenty-first century's Reichstag is even green and high-tech. But it's not merely all those fine things. It's also cheap. And the budget is never a matter that a statesman overlooks—as the next chapter points out in detail.

THE PANTALOON

> *The sixth age shifts*
> *Into the lean and slipper'd pantaloon,*
> *With spectacles on nose and pouch on side.*

This is the chapter about getting rich. The businesslike, no-nonsense chapter. Let's get straight to the bottom line. How about some nice big piles of money?

A lot of people write books about how to get rich. There's a steady, predictable, even ferocious demand for these books. Since one of my hobbies is corporate futurism, I know quite a few of these business authors. Here in stage six I've finally become one of them. This is the part where I tell you how to become wealthy beyond your wildest dreams.

First, however, we should try to be honest with each other. We should boldly confront a major, high-concept problem in the business-consultancy biz—namely, "Hey! If this guy's so smart, why isn't *he* rich?"

After all (the reader should properly reason), if *I* were rich, I sure wouldn't be busting my hindquarters writing business books. I'd be doing the things I always imagined rich people doing, such as hanging out in a palatial hotel suite in a tropical paradise. I certainly wouldn't be burning the midnight oil teaching other people how to make money.

Because what on earth is the point of that? Let's take a hard, stern look at the real-world economics of writing a business-advice book. Our author, being a mere author, is making only 6 to 8 percent royalties out of the product's revenue stream. That means that 92 percent of his effort is going straight to publishers, distributors, and bookstores. If he really understands business as he claims, how come he's writing a book? Shouldn't he cut to the chase and become a publisher? Consider a shining role model like William Randolph Hearst, the publishing mogul who built the fabulous palace of San Simeon and was the inspiration for *Citizen Kane.* Now, that guy was *rich.*

So I guess I've rather undercut my own credibility here, but bear with me, because I'm about to make a compelling case that I actually *am* rich. Plenty rich. Therefore, you should pay attention to me. You should listen with rapt fascination as I describe the future of business, even though I've never done a lick of work in the business world and I have absolutely no intention of soiling my authorly self with the plebeian hurly-burly of mere commerce.

What about that bottom line, eh? Exactly how rich am I? Well, I'm not the richest guy in the world. Before their stocks tanked, that would have been a toss-up between Bill Gates and Larry Ellison. Tragically, I'm not even the richest author in the world. That would be Stephen King. When we fiction authors are talking business up in some hotel suite, spilling Chardonnay on our shoes, Stephen King is our industry touchstone. "Oh, yeah," we say feverishly, "if only *I* could be Stephen King! Haunted by horrific visions, losing my eyesight, hospitalized by a runaway truck—man, that's the life for me!"

But although I'm not Stephen King, I'm still pretty rich. Looking at the tax statistics when April 15 rolls around, I can see, with wide-eyed incredulity, that I command more revenue and resources than 99 percent of this planet's population. So even though I dress like a permanent grad student and I live in cyberpunk squalor, there's no question that I am a member of the planet's financial elite. I've even been to the Davos World Economic Forum. And I didn't have to pay. I was *invited faculty*.

I know that this must seem like a shocking confession. If I were to declare that I was bisexual, or an alcoholic, or a manic-depressive, I'd be swiftly forgiven; after all, I'm an author. Whereas gloating publicly over one's bloated income stream is really abrasive and appallingly déclassé. For the people whom David Brooks calls "bourgeois bohemians," bragging about money is a dead-serious social taboo. If I were to cut and paste my latest 1040 tax form onto the page here, it would be far worse and more shocking than posting nude pics of myself on the Internet.

Nevertheless, here in stage six, I'm determined to come clean on the subject of money. Because it's healing and empowering to confront these ugly truths about ourselves. It helped my mental health a lot when I finally broke down, sobbing bitterly over my platinum credit cards, and admitted that I had somehow become a rich guy. It's really no use being all evasive and dainty about this. The time comes when denial is worse than the truth. Therefore, I am bound and determined to testify about the peculiar nature of the modern and future generation of wealth, and to move from the abstract to the specific, I'll make a case study of myself.

When I first began writing books twenty-five years ago, back in the Legendary Early Days of Bitter Struggle, I was young and sloppy and bohemian, but I wasn't truly poor. Though I wore ragged tennis shoes and lived mostly off canned chili, I was having a pretty good time. My dad was a well-to-do professional who had put me through four years of college. I had a day job. Even my charming and delightful live-in girlfriend was employed and pulling her own weight, as well as, if truth be told, some of mine.

I was taking a risk by choosing the writing life, because I was writing science fiction novels, an activity that would lie like a lump of putty on any future business résumé. But I could afford to experiment. As it happened, I knew what real poverty looked and smelled like, because as a teenager I had lived in India. It's sometimes hard to understand that you are privileged, but true, abject destitution is unforgettable. Unlike prosperity, destitution can never be mistaken for anything else.

My corporate-futurist colleagues commonly use scenarios to confront us with the consequences of our actions. Handled properly, they can be very revealing; a good scenario will slice through layers of time like a cake knife. So let's imagine a futurist scenario about my personal economic situation.

Imagine that I took my younger self, the twenty-year-old wanna-be novelist, and brought him here, into the futuristic present. We'll escort this jumpy youngster into my office here, tomorrow now, the dodgy patch of space-time where I happen to be typing this book.

There is no avoiding the conclusion that this young man would be stunned. He could have moved all his earthly possessions into the space where I shelve my novels. It's a rather nice office—not that this would impress him. Like most bohemian children of prosperity, he had a fine contempt for mere possessions. No doubt he'd make a beeline for those novels I wrote and start thumbing through them, wondering where I had gone soft, sold out, and lost my edge. That kid was a blazing fanatic who desperately wanted to be heard. It never occurred to him that his strange thoughts would bring him money. But that happened anyway. He wasn't stoked for that part. That was genuinely surprising and alarming.

If money happens, people have to deal with that event. Even if the getting-rich part is easy—you might win a lottery—the consequences are not. There are many books written about how to get rich. There are very few written about how to be rich. Being rich is basically a folk handicraft. It is the lifestyle of a tiny, eccentric minority. When not making money (their central passion), rich people spend their free time watching one another for cues on how to behave properly. They also spend a lot of time trying to avoid being cruelly ambushed by hucksters who want their cash.

If you come to know serious rich people, say at the World Economic Forum or the Renaissance Weekend, you'll notice that they always make that peculiar wincing motion with their lips when the name of Donald Trump pops up. Donald Trump is very rich, but he is the chrome-plated cartoon version of a rich guy. He deliberately and loudly does the things poor people imagine rich people doing, like building a yacht, picking up chicks, and running for president. Most rich people are shy woodland creatures who are totally not down with that. Genuinely rich people tend to be rather reserved and courteous, because quiet courtesy is the most efficient way to get rid of people who want your money. They don't do rich-guy cool things or rich-guy fun things.

That's because fun rich-guy things cannot make you rich—they make *other people* rich. A person who joyfully splurges on wine, women, and song is methodically moving money into the hands of ex-wives, lawyers, and interior decorators. He's not getting rich; he's getting poor.

True fortune-builders do not pursue this form of behavior. It is radically alien to their temperament. They use their capital, but they don't splurge. A fortune is not a vast inert mass, like the dollar-sign bag in Monopoly, but rather an ongoing process. True wealth is about compound interest and return on investment, which means placing capital only in areas of endeavor that will create and supply *more* wealth. For the core fortune-builder, the captain of industry, this is the whole point. Under no circumstances will he knock it off with the wealth building in order to "relax and enjoy it." Nobody needs two hundred million dollars in order to relax and enjoy themselves. The only reason to have two hundred million is to transform it into two billion. Once you've got two billion, then you can start getting *serious*.

If you read the memoirs of the very wealthy, you will never find them chuckling to themselves about how they pulled a lot of cash out of the bank and fed champagne to their racehorses. They may do this sort of thing, but only because their wife or husband insisted on it and brought some people with cameras. The very wealthy tend to be tightly focused, lugubrious workaholics. Their books reveal no amazing secrets to wealth. They never talk about Ponzi pyramids, surefire investment schemes, gold mines, or government loopholes. These lame fraud schemes are mere come-ons, trading on the greed of gullible people who don't get it about capitalism.

Even if some long-shot lottery scheme delivers a truckload of cash to you, you can't keep that money unless you successfully manage the consequences of having it.

This is where I, a mere author, part company from people who are seriously rich. I'm not a capitalist. I am just a well-rewarded independent contractor in the culture industry, an entertainer, and my fickle prosperity could vanish at the public's frown. I turn words into publishers' advances, but I've never managed to turn money into more money. I've seen it done, and I know how it's done, but I don't do that.

The reason that I don't do it—at least not yet—is because it's a big, stuffy drag. Oddly, it really is that simple. Shakespeare's shrunken pantaloon makes a perfect rich guy because he's patient, he's persistent, he can really focus on detail, yet he lacks joie de vivre. Just look at him: he's got spectacles and skinny legs; he's kinda wrinkly, clammy, and geeky. It's a chore to kiss this guy; women who marry him for his money are going to earn it the hard way.

There's a stiff price to pay for taking money seriously. The price is that I have only one life: a finite number of usable human hours. When I carelessly throw my money out the door, instead of salting it away in CDs, REIs, Treasury bonds, or mutual funds, it probably looks crazy, but that's only half my story. When I do something unlikely and extravagant, like, for instance, hanging out in Munich at a conference on brain transplantation, I'm not merely damaging my financial future—I'm also *becoming a better science fiction writer.* I'm soberly investing in my own skill set as an independent contractor. Because, man, that material in Munich was fantastic! You should have seen this professor's white paper on using robots to explore the East German sewers. Then I had lunch with this Australian performance artist who pierces his flesh with steel hooks and has himself rolled out of windows on a clothesline . . . but I digress. I suppose it's in my very nature to digress. In business, however, digression is generally fatal.

After I finished sobbing bitterly over the fact that my books were selling, I somehow came to terms with having some money. Capitalism and I reached a stable, if not particularly fruitful, accommodation. It works like this. I write really weird books, and some money arrives. Then I go spend the money in weird places and learn about weird stuff. My books get weirder yet. In theory, this process could go on quite smoothly until I can't type anymore or no one wants to pay.

But alas, time passes, the old verities change. The strangest thing about my relationship to capitalism is how close the business world has moved to science fiction. As years passed and my career advanced, business moved with increasing speed and aggression into my own cultural territory. Science fiction has always been a byword for the weird and improbable, but the turn of the century marked the first time in which I began receiving serious job offers in business. Businesspeople began asking me to take executive posts, to join advisory committees and corporate boards of directors.

It was no use my pointing out to them that I had never met a payroll in my life, had no executive experience, and was utterly disinterested in meeting stockholder expectations. They knew all that already. In fact, they *liked* that part. That was why they were seeking me out. I earn my living making up weird imaginary crap, and they considered this a major business asset.

"That's just why we *need you*," they would insist, with pained rich-guy expressions. And it wasn't just dot-com cranks, whose behavior is notoriously manic and elated. *The Wall Street Journal* began commissioning opinion-editorial pieces from me. I found myself writing for *Fortune* magazine.

I was somewhat disturbed by this. Once I even asked for advice and counsel at a Washington committee meeting of the Rockefeller Foundation and the National Academy of Sciences. These august experts (I figured) ought to know why this was happening to me. But nobody let me off the hook, and in fact an Ivy League professor publicly accused me of being "disingenuous." *Why*, she demanded, was I trying to downplay my influence on technocommercial development?

I found this peculiar. I'm rather interested in businesspeople as a social class, but no more than I am in, say, politicians or the military. Politicians never ran me for office, and the military never drafted me. So why was business so hot to trot all of a sudden?

If this was madness, it was all over my town. In 1975 my hometown of Austin, Texas, was full of ragged-jeans, cheery, arty, downmarket types, just like my younger self. By the dawn of the twenty-first century, it was full of designer-jeans, cheery, businesslike, upmarket types, like those people who wanted to hire me. It's not that we Austinites slowly lost our counterculture hipster values and sold out to the Man, baby. That wouldn't be news. That's been going on in bohemian circles for 150 solid years, and the game is well understood. All the graying fat cats would clear out for parenthood and the suburbs, and Austin's bohemian underground would fill up instantly with the next wave of college students. Counterculture as usual, no problem there at all.

This transformation was different, more fundamental, more serious. Somehow, society had decided to commodify intellectual property, to make vaporous rantings worth cold, hard money. What used to be sophomoric philosophizing around a pizza and a pitcher suddenly became a valuable exchange of marketable information. Even the little stuff was getting soaked up. Job chats became job-hunting websites. Phone chats became catalogable e-mail. Garage sales became eBay.

Why was this? Because there really is a new economy—not necessarily better, but one that's new in character. An information economy has a different technical infrastructure and new methods of generating (and losing) wealth. You can see this in simple things, like business purchases of office equipment. There were some computers around in 1960; they were usually built and sold by International Business Machines, and they took up about 15 percent of big-business outlays on equipment. That proportion became 25 percent by 1980, at about the time computers and networks got to be called "technology," just as if all other technologies had ceased to have relevance. In 2000 business-equipment spending was 53 percent "technology." Businesses are what they use; these are the tools of the trade.

Businesspeople don't buy computers just because they blink nicely and make cool beeping sounds. They're bought for business reasons: to shave time and money off manu-facturing processes and distribution; to compile massive databases on customers and suppliers. They improve com-petitive advantages, they open up new kinds of market-places, they coordinate just-in-time shipping, they turn customers into end users, and they pave the Earth in com-mercial blobjects. "Information" has changed—because it got economized. The blithering, brainy stuff that was once just a steamy cultural haze can now be slotted, stored, recorded, and turned into files and algorithms, placed on backup disks, and whipped through networks.

It's not just Austin that turned from a sleepy college town into a software and Web-content countinghouse. Much the same fate has befallen all of Austin's sister cities in America, such as Berkeley, Burlington, Charlotte, Chattanooga, Madison, Portland, Raleigh, and San Diego. These once modest places were college towns and government towns. Colleges are the preindustrial institutions who make it their business to spew information. Governments breed vast ranks of clerks and data centers, for government is the historical source of statistics and records. College and government allowed the literate to survive, but computer networks made the literate into the computer-literate; that made them postindustrial mavens. So places once ideally suited to bookish wanna-be novelists are now ideally suited to UNIX gurus.

Now let me explain to you how it is that people create and sell information goods. If you want to become a twenty-first-century mogul, this knowledge will be of vital use. Whereas if you are merely a clever, creative person who is bored to death by commerce of any kind, this briefing may keep you from getting all demoralized and confused when it suddenly dawns on you that you are somehow making complete strangers really rich.

To understand the contemporary information business, we have to dissect a core idea, which is that "information wants to be free." This famous aphorism was coined by Stewart Brand, a truly great American who is both a global corporate futurist and a hard-core hippie freak. Although Brand is not profoundly wealthy and has never been elected to any public office, he is clearly one of the most influential people in the whole world. This can be proved by the fact that, although he has scarcely changed a bit since his salad days in the 1950s, the world at the turn of the century looked very much like Stewart Brand has always looked.

It's not just that he is both hip and high-tech, a congenial Californian carrying a top-end digital camera and wearing remarkably sturdy shoes. Almost everybody does that nowadays, whereas Stewart was once a total anomaly. He was a man with a double life, a hard-core activist Summer of Love California hipster who nevertheless had serious, prolonged, profound engagements with giant multinational oil companies. His behavior was so contradictory and inexplicable that it couldn't register with people. Their eyes just slid right off him in an inexplicable panic, as if he were Aleister Crowley.

By the year 2000, however, the planet was financially and technologically dominated by people who spoke and thought in fluent Stewart Brand–ese, even if they didn't know this. The only real difference between them and him is that

Stewart was pushing seventy, while they were pushing thirty and considered themselves bang up-to-date.

The secret sting in Stewart's famous aphorism "Information wants to be free" is in the part that got clipped off when the slogan became popularized. The full quote ran: "Information wants to be free; information also wants to be expensive." It's only when you can hold both of these principles in your head at once that you become a true information-economy adept. If you focus exclusively on one half or the other, you will spend most of your time profoundly lost, in a thrashing state of hapless resentment.

Of course (as baffled, literal-minded engineers never tire of pointing out), information doesn't "want" anything. This statement is a metaphor, encapsulating a trend. Information wants to be free for two very good reasons: (1) the techniques for storing it and hauling it around have broken loose from most of the previous physical constraints; and (2) eventually, you just can't get people to pay you for it.

Exhibit number one is William Shakespeare. Shakespeare did okay financially, but he never made all that much money, and he's certainly not making any money now. If you want some Shakespeare, you can just log on and get all you want. People rip off Shakespeare every day. Shakespeare never gets a dime. End of story.

The value system attached to information is cultural. When culture changes, then all the useless stuff in the attic becomes the public domain. Not only is it free then, but as with Windows 1.0 or the Apple II, we'll pay to get rid of it. Dead technology and dead information are just as dead as any other form of dead culture—in fact, they're even deader. Information systems require huge teams of people and a steadily ongoing effort to maintain them. Otherwise, they die and rot and cannot be resuscitated without fabulous expense.

It's pretty easy to make a 1945-style gun or helmet, but if you want to re-create a 1945-style Enigma computer (one of the very simplest), you've got a lot of work on your hands. It's not just the chips and wiring that are tough to make and complex to operate; it's information per se. Egyptian hiero-glyphics were made with mere rock and stone, ink and papyrus, yet when people forgot how to use that informa-tion system, hieroglyphics remained in plain sight, com-pletely impenetrable and utterly mysterious, for centuries.

Information wants be be expensive because, even though a stream of ones and zeros seems cheap and easy to store and reproduce, digital data can exist only in an elaborate tech-nosocial context. Therefore, there are any number of clever ancillary ways that I can spin it, stack it, or spill it to get some kind of legal, cultural, or industrial hammerlock on you. If I have your eye and ear, then I also have your nose. It's not about "friction-free" ones and zeros, whipping gaily around cyberspace at the speed of light. Who cares about ones and zeros, anyway? Ones and zeros don't have checking accounts and can't pay anybody a thing. Information business is all about me, the vendor, using information to maneuver you, the buyer, into a situation where you have no recourse but to give me some cash.

As our basic business model, we'll use novel writing. That's a very old-economy, primeval information business. I make some money at it, but this activity doesn't have to be strictly commodified. Science fiction also wants to be free. If you want to be entertained by a science fiction writer without paying him, you can choose Jules Verne, a nice dead one.

Or maybe you can show up at one of my house parties and hang out here in the much-cited office where I'm typing this book. There you'll see me being all finger-snappingly witty, and spewing peculiar ideas, and showing off bizarre videos, and forcing weird art magazines on innocent people, and loudly name-dropping my acquaintanceship with True Gurus such as Stewart Brand. You can even have some free taco chips and a beer.

People who have undergone this strictly noncommercial process often claim that they understand my novels a lot better. The crimp here is that although a party at my house has plenty of free information, it's not an industrial, commercial process. It's not scalable, consistent, and repeatable. A Bruce Sterling book like this one has been heroically edited, and proofread, and printed, and shipped all around in a consistent edition. At a party I don't deliver the same talk to everybody. In fact, the quality varies dramatically.

So even though I'm shaking your hand and breathing on you, there's a lot of noise and repetition at these gigs. Toward 2 A.M. when we're generally all pretty wasted, you'll be getting a seriously degraded version of the experience.

And what if I've wandered off willy-nilly into some passionate rant about some hobbyhorse of mine, such as computer search-and-seizure practices? Ooh, that could get really bad—because since you're not paying, you don't get to pick the tune. You're in a completely authentic situation, entirely divorced from the sordid demands of mere commerce. Therefore, you'll actually hear what's on my mind, as opposed to what some market demographic is willing to pay for. People who think there is no downside to this should talk to the *wives* and *husbands* of authors.

So here we have a first vital trick: my publishers don't just sell you information. They *spare* you information. Publishers are speculators, intermediaries, and editors, as well as locators, filters, promoters, and stockpilers. All of this adds value to the raw-information stream that I spew through some congenital psychological defect. This book is a commercial product, so it's manageable, it's filtered, it's convenient, it makes few demands. You can pick it up, put it down, throw it across the room, sell it to someone else. None of those handy things apply to the author.

Authors spew information like live volcanoes, but you don't *want* a free author in your house. Authors are very troublesome; they are egotistical and demanding, and commonly tipsy, or bipolar, or undergoing a loud and painful divorce. Publishers, by contrast, are more or less okay, except for their periodic crazy attacks of mergers and downsizing. Publishers can *insulate* and *protect* you from the free information emitted by authors. Without their paid services, it would be very much as if you'd married an author.

Books look the way they do because of the second useful stunt in information economics: "versioning." Like most methods of making cash from information, versions are akin to psychological warfare. Versions wring extra money out of the sales process by dividing the market into vulnerable demographic chunks whose arms can be twisted in clever computer-targeted ways.

In publishing's days of yore, this used to be about expensive, classy hardbacks (for rich people) and cheap, cruddy paperbacks (for poor people). But no more. The situation as it stands today is far more complex.

Step one in modern publishing is the Advanced Reading Copy Not for Sale. These promotional versions are aimed at reviewers, chain-store reps, and other literary tastemakers. These cost *nothing;* they show up *for free.* The publisher gets no money for it, and yet it can be a sizable edition, often several thousand. I get a lot of these freebies. Yet I do not dance and exult when I get them. For those of us heaving and whining at the star-maker machinery, they are something of a burden.

Then there's the boxed collector small-press edition,

with gilt leaf and ribbon, brought out by a small-press ally and microtargeted at wealthy fan completists. These books cost a small fortune, yet they tend to sell out fast. That is because these customers are not merely enthusiasts, they are a book market in and of themselves. They not only buy these collector versions, they refuse to read them, because they carefully wrap them in Mylar. Years later they will sell them at steep markups.

Then there's a leather-bound edition, a mere handful of copies, for just the author and some of the editors and office hangers-on. These gift books cost nothing but are worth plenty, especially after the author is dead, for then they become a precious "associational copy" onto which the dead author's magic spiritual aura has somehow rubbed off.

Only then does a proper dust-jacketed hardback version appear, and then all the straight people in the chain stores snap it up, fondly imagining it to be the "first edition."

Months later the trade paperback version shows up, for folks who don't buy pricey hardbacks yet don't want to be reduced to the cheap stuff. In marketing psychology, this buying pattern is known as "extremeness aversion," and we'll get back to it later.

A year later the mass-market paperback hits the market. This used to be the huge edition, the mainstay of bus depots, drugstores, and train stations. Often this paperback step is omitted entirely nowadays. Although a mass-market paperback still sells a lot of copies, the margins are too thin. Also, there's a serious rubbish problem with paperbacks. They occupy valuable retail rack space, to the detriment of other goods flowing through, and it costs too much to get rid of them if they don't sell in a hurry.

Then comes the remainder table (again, an industry in itself). This discounted version is the last actual "version" from which the publisher and author can make any money.

But that's by no means the end of the story, for the secondhand book trade is yet another entire industry, in which mere bozos like authors have no financial interest. Above this vast garage flea market is the charmed, angelic circle of the bibliophile trade, where antique books with no commercial appeal will nevertheless command huge prices from eccentrics who, for whatever reason, really want them around.

The literary demimonde turns full circle when the critics who get the Advanced Reading Copy Not for Sale take it directly to the secondhand guys. These book culturati are receiving a kind of off-the-books subsidy. This tactful payola gently encourages them to spend more of their valuable time reading books, and so acting as unpaid promotional agents in the print-based culture industry.

The funny thing about publishing as an information industry is that although everyone involved is always in despair of it, it's very highly evolved. It's a mature industry with a great deal of history, where all the players have been reduced to near-charity or shabby genteel nickel-and-diming. Mature industries tend to reach that state; if the means of production and distribution are well understood and easy for competitors to replicate, then there is little volatility, and thus little chance to receive large rewards for small efforts.

For years on end the computer industry behaved in that volatile fashion. Only exploding, unstable information industries like software and Web commerce circa 1999 have fantastic rates of financial return. In those halcyon days, nobody knew where the industry was going, and a kid in a garage with a big idea and some patents might become the next Croesus.

Eventually, when the remaining hype and novelty wear off the information economy, these new industries should either cease to exist or come to look more or less like book publishing. But it's not at all likely that book publishing will come to look more or less like them. The landscape's just too well plowed in those parts.

With the ferocious triple dominance of Microsoft in operating systems, Intel in chips, and Dell in hardware, the computer industry finally began to look a bit boring. Almost as boring as books, and in some interesting ways. This computer oligopoly still pretends to innovate for form's sake, and to scare off the trustbusters; but the glamour routine is becoming ritualized. The machines are slow, the programs are bloated, the changes are cosmetic—very much like the planned obsolescence of Detroit's Big Three carmakers in their heyday.

Computers, software, and Web commerce have long treasured their profound newfangled mechanical advantage over book publishing. Their hardware is unstable and expensive—and that's not by accident, either.

Watching book publishing is like watching a centuries-old cocktail party where academics and pornographers spill Chablis on each other. But the computer business wants to be really hot. Watching the computer business is like eavesdropping on a rich kid's affair with a bipolar supermodel. He (the user) is eager, he's gullible, he'll fall for anything, but he is lethally promiscuous and fickle; if she's anything less than taut, hot, and totally glittering, it's right off the edge of the loading dock, baby. She (the vendor) is a lean and mean beanpole-tall jet-setter who's always heaving iron in her gym or preening before the cameras, while screaming hysterically for next season's clothes, now, right away, tomorrow now. As long as both of them don't know what's coming next, they'll be glamorous as all get-out.

In the computer world, the user plays stage-door johnny, while next year's Intel Pentium running next year's Windows is the shrill, abusive, and demanding arm candy. This theatrical concept, while not very flattering and somewhat sexist and demeaning, explains a host of phenomena about the information economy. Because the information economy is not about the information or the economy. Everything important that happens there is about the *relationship*.

The information economy is about who promises what to whom.

In the information business, it's not really about who's fastest, most advanced, or most high-tech; that's just the sexy croon of the industry's come-on. Behind the scenes, it's all about commitment. The point is to make it harder and more wounding to break up with me, the vendor, than it is to put up with my continual exploitation. There are basically six ways to do this, and all of them are used in the information business, all of the time.

1. A contract. We'll put it on paper, we'll make it legal. I'm using you and you're using me, but I'm not leaving and neither are you. Somehow we've concluded that we really need each other in order to go on living. We stand in front of witnesses, and we agree to stick it out together somehow. It's normal, it's honest, it works. Unless it doesn't work, in which case it gets extremely ugly and leaves permanent emotional scars.

2. Brand-specific training. I'm really complicated and hard to figure out, but I give you something that you just can't seem to get from anyone else. We've spent thousands of hours talking over every single little quirk and kink I have, and getting completely wrapped up in my needs. Once you've gone to the incredible trouble to fully understand me, it just seems totally exhausting to start all over again with somebody new.

3. Search costs. There's probably someone else who would suit you as well as I do, but thanks to circumstances (which I've likely helped to arrange), you'll probably never find them. Not around a sorry little town like this, anyhow. (Search costs used to work really well in the days before the Internet. Now they're history. This has created an almighty panic among vendors, who now know that users can ditch them without mercy for some distant floozy in Tokyo or Toronto.)

4. Information formats. No one else can even speak our language around here. Our private little argot involves arcana like dead computer languages and voodoo keyboard rituals and huge product catalogs, all written in our own private lovers' baby talk. If you ditch me and try to pick up somebody else by talking that way, she'll look at you like you came straight from Mars.

5. Durable purchases. You bought a huge mainframe, and a specialized numerical milling machine, and special scanners and printers, and a car and a fridge and a sink. You're not just going to *walk away* from all of that, are you? Boy, can I ever make that cost you.

6. Loyalty programs. The more you hang out with me, the more I seem to like you. In fact, I'm always coming up with sweet little favors, based on how well we're getting to know each other. I can get you a window seat, and upgrade you to first-class, and throw in a free Web browser with my operating system. In fact, why not bring over your friends and family? Since I'm so thoughtful and generous, your mom and

dad are sure to love me, too. Then I'll be especially sweet to you, and I can whip up that big turkey dinner you like, and we'll all get real cozy.

There are many other aspects of informational coquetry that may not seem very "technical," yet they still achieve a commercial end. These gambits all get folks to pay big expensive wads of money for some data that "wants to be free."

A. Branding and reputation. Look, babe, you can trust me. My famous family of products has been around for years. I'm respectable, decent, and dependable; I'm really not that kind of guy. You've known me since you were a kid; why would I risk all that just to take advantage of you in this one little situation? Stick with me and save yourself a lot of worry and trouble; if you liked my newspaper and TV show, you'll love my broadband cable.

B. Standard setting. Everybody depends on me. I shoulder the grave responsibility of being reliable, dependable, and thoroughly predictable. I am the authoritative source through which all good things flow. If it doesn't work with my stuff, it just plain doesn't work. International committees say so. And if I get my way, so will the government.

C. Expectations management. (Also known as FUD or "fear, uncertainty, and doubt.") I know you're thinking of buying from that other vendor, but you'd better not. His stuff is half-baked and will injure you. Besides, I'm making one of those myself, just next quarter. Mine will be much better than his, so everybody's sure to use mine. If you insolently dare to buy that lousy one, everyone will laugh at you, and you'll just have to buy it all over again from me, anyway. And you'd better believe this is no idle threat, either, because the woods are full of upstarts that I stepped on.

D. Creeping featuritis. This common but unpleasant practice adds more and more "attractive features" to a product, merely to keep the jaded user intrigued. You like mascara? Lip gloss? Piercings? A mascara that's also a lip gloss? How about some clown wigs? No, look, look, I just invented a whole pull-down menu full of rubber noses!

E. Sell the organization, not the bit stream. Let's redefine our relationship. You're not buying any mere "information" from me. Forget that part; we both know it's just tangential, not really serious. You are *hiring me*, a valued adviser, a grand vizier, an indispensable part of your management team. I possess profound cybernetic insight, and I know your wants and needs better than you do yourself. In fact, why not make me your prime minister? Just give me your checkbook, and I'll take care of the rest.

F. Dubbed local versions. It's way too crowded and sweaty here in the English-language ballroom; I'll go off and vamp in Russian, Chinese, and Hindi.

G. Markets of one. Specialized local versions of products have been osmotic cultural forces throughout history and are one of the great driving forces of civilization. However, computers and networks allow this exploration of markets to be carried to unheard-of levels of microtargeted sophistication. This is truly the great breakthrough of e-commerce, the thing that has professional marketers really excited. It's the ultimate relationship move: a profoundly new ability to split entire nations into markets of single individuals. Handle this right, finesse the legal and privacy problems, and you can compile dossiers on customers that would have dazzled the gestapo. My whole organization can now single you out, study your every least nod and blink, and gang up on you to stunning effect. It looks like this.

I know you've seen an old-economy brick-and-mortar bicycle. But this bicycle is the Revolution, man. Because this is no mere bicycle, this is *your* bicycle, which we like to call "MyBicycle." MyBicycle doesn't have our brand name on it—it has *your* name on it! Wow!

No one is comfortable on that bicycle seat but you. The pedals are your favorite color, and they fit only your feet. Look at this rich array of stickers and decals we've provided, with which you yourself can invest your own individual creativity into MyBicycle. Not only that, but MyBicycle knows your continent, your country, your state, your town, and your neighborhood map. So MyBicycle knows when you're going for bread and milk. MyBicycle knows how much change you have in your bicycle shorts. MyBicycle even drives more slowly whenever you're drunk.

And whenever you put something in MyBicycle's cargo basket, we always know.

You needn't worry about prices, either, because when you're riding MyBicycle, all the prices are automatically arranged just for you and you alone! If you're in a good mood and in a hurry, prices are high! If you're feeling chintzy or suspicious, they're lower. Best of all, if you fall off MyBicycle and break your leg, you can't sue us, because hey, that's *your* bicycle. You built it yourself, not us!

So that is the story, ladies and gentlemen. At least, those are the standard methods that we've cooked up in the information economy's early days. We've come quite a long way from the notion that "information wants to be free," and by golly, we've somehow been paying all the way! All that money changing hands, despite certain dead-obvious facts about information goods.

Information has large up-front sunk costs. Let me tell you what that means in English. It costs a fortune to make the first copy of that information, just like it somehow took me a whole year to hammer away on this slender little book. All that time, the publisher was sweating bullets lest a concrete block hit me in the head and cost them a generous advance.

But once the manuscript makes it safely down to the presses, it's an entirely different ball game. Then, instead of stumbling around in some limbo of poorly disciplined, artisanal creativity, this book will spit out of the printing machines like nobody's business. Copy number 10,112 costs just the same as copy number 00002.

Information goods have minimal capacity constraints and low incremental costs. In other words, the vendor can print as many books as it wants to and is not at all likely to suddenly run out of ink. It can go back and print more books any time it pleases. Of course, publishers are hampered by an ancient, goofy distribution system. But if they could leverage the Internet in their own favor, they'd find that their distribution costs had also collapsed. That would mean Oz. Publishers would have a marvelous product that had once cost them a lot of money to make but would spew money back into their pockets, indefinitely.

At this point the "information wants to be free" enthusiast horns in from the radical fringes of cyberspace. And it turns out that he can underprice the works, because he didn't have to pay to make the thing. All he has to do is copy it, which is supercheap and easy. He loudly asserts that *he, too,* can copy and distribute that book at little or no added cost! Yay! Piracy rulez, dood! Thanks to my computer and modem, I've short-circuited the whole evil capitalist charade, which is now ready to collapse like a pack of cards! It's a simple matter of scanning the book (in a handsome PDF format, say) and sticking it onto a website! Then all I have to do is place it offshore in some ninja-haunted concrete data haven and defy the police to come get me!

But to say, with real though rhetorical force, that information wants to be free is not at all to say that information is actually free. Nobody has ever said that. Friction in the marketplace does not vanish entirely just because ink becomes ones and zeros. It merely mutates into new forms of friction.

Bits, for instance, are not immaterial. Bits in motion are physical things, electrons and photons. Bits sitting still are little patches of magnetized metal flakes, or little black pits of molten plastic. They're real objects, bits of atoms. They *seem* immaterial compared with wood pulp or tombstones, but if you visit a modern Internet backbone router, you will find yourself in a very large, extremely material info factory that sucks voltage just like a steel mill. A backbone router doesn't employ any working-class guys in blue overalls, but it definitely occupies space and has mass; it has plumbing and pays real-estate taxes.

And it's delicate. The least little interruption in current for one of these e-commerce refineries will provoke a devastating financial meltdown. So these supposedly ethereal Internet nodes are commonly built with big, hairy backup diesel generators on-site. These cybernetic smokestacks are every bit as solid and terrifying as any of William Blake's "dark satanic mills."

All machines that store, manipulate, or move bits are cranky and temperamental. In particular, personal computers have long been deliberately designed for replacement every eighteen months. Computers die quick and are meant to die quick. Their quick death is in the financial interest not just of the people who make them but of the managerial and programming classes that use them.

If computers lasted forever and were simple to use, the high-paying jobs in e-commerce would immediately migrate to India. In fact, they are already doing just that, at an impressive clip. India is not yet a great center of radically destabilizing innovation, but it has oodles of literate people who can read manuals and push buttons. The more slowly computers evolve, the faster the salaries shift, away from expensive geeks and off to anonymous clerks.

This friability has grave consequences for information economics. In fact, it shapes the whole business. It's not enough to invent new ways of using computers; for maximum profit, the old ones must be killed.

The local university has a Gutenberg Bible that is perfectly legible. It was one of the first books ever printed, and though it's worth about two million dollars now, it's still the same book it was when Gutenberg was publishing it (and slowly going broke).

By stark contrast, consider this elephant's graveyard of personal computers:

Altair 8800, Amiga 500, Amiga 1000, Amstrad, Apple I, II, II+, IIc, IIe, IIGS, III, Apple Lisa, Apple Lisa MacXL, Apricot, Atari 400 and 800 XL, XE, ST, Atari 800XL, Atari 1200XL, Atari XE, Basis 190, BBC Micro,

Bondwell 2, Cambridge Z-88, Canon Cat, Columbia
Portable, Commodore C64, Commodore Vic-20, Com-
modore Plus 4, Commodore Pet, Commodore 128,
CompuPro "Big 16," Cromemco Z-2D, Cromemco
Dazzler, Cromemco System 3, DEC Rainbow, DOT
Portable, Eagle II, Dragon System Dragon 32 and
Dragon 64, Epson QX-10, Epson HX-20, Epson PX-8
Geneva, Exidy Sorcerer, Franklin Ace 500, Franklin
Ace 1200, Fujitsu Bubcom 80, Gavilan, Grid Compass,
Heath/Zenith, Hitachi Peach, Hyperion, IBM PC 640K,
IBM XT, IBM Portable, IBM PCjr, IMSAI 8080, In-
telligent Systems Compucolor and Intecolor, Intertek
Superbrain II, Ithaca Intersystems DPS-1, Kaypro 2x,
Linus WriteTop, Mac 128, 512, 512KE, Mattel Aquar-
ius, Micro-Professor MPF-II, Morrow MicroDecision 3,
Morrow Portable, NEC PC-8081, NEC Starlet 8401-LS,
NEC 8201A Portable, NEC 8401A, NorthStar Advan-
tage, NorthStar Horizon, Ohio Scientific, Oric, Osborne
1, Osborne Executive, Panasonic, Sanyo 1255, Sanyo PC
1250, Sinclair ZX-80, Sinclair ZX-81, Sinclair Spectrum, Sol Model 20, Sony SMC-70, Spectravideo SV-
328, Tandy 1000, Tandy 1000SL, Tandy Coco 1, Tandy
Coco 2, Tandy Coco 3, TRS-80 models I, II, III, IV, 100,
Tano Dragon, TI 99/4, Timex/Sinclair 1000, Timex/Sin-
clair color computer, TRW/Fujitsu 3450, Vector 4, Victor 9000, Workslate Xerox 820 II, Xerox Alto, Xerox
Dorado, Xerox 1108, Yamaha CX5M.

That's merely a tattered, very partial list of dead computers. And here's the kicker: think of all the *free information* those computers had inside them when they perished! *Free* information? It *wanted* to be free; in reality people paid to be rid of it.

Web pages die much faster than personal computers do. The average life span for a Web page on the Internet is about forty days. What does this mean? It means that keeping "free information" alive and available is a great deal of serious work. Unless some income stream pops up from somewhere, that work goes unrewarded. This means that people get discouraged and cease doing it.

If the enthusiast counters that there are other means of reward for promulgating, archiving, preserving, and distributing free information, then the argument has shifted. He is no longer claiming that information is free. He is merely championing some alternative economic arrangement. Something not profit-centric. There have always been plenty of those. Libraries and governmental archives are supported by taxes. Universities and academies are supported by tuition and land grants. Clubs are supported by dues and bake sales. That doesn't make them unimportant or not worth doing. But it doesn't make them free, either.

A pirate can sell information goods, but if he sells too many, he gets too obvious. It's hard to be both famous and obscure; that's how terrorists, warlords, and mafiosi live, and it takes a lot of hard work. If pirates don't sell enough goods, they're wasting their time. If they do make money, then they themselves have to worry about piracy—they must fear the lawless rival pirates who could underprice them or cheerfully shoot them.

This is not to say that the status quo is perfect or even stable. There are certain aspects of commodifying information that are frankly loathsome. They're legal, they're profitable, but they say things about human psychology that are very unflattering. For instance: did you know that software manufacturers create their top-of-the-line version first, and then deliberately cripple the cheaper versions? They do this in order to be sure that you'll get less than your money's worth.

Think about this, knowing what one does about sunk costs. For precisely the same effort it takes to sell you junk, they could have put their very best ones and zeros on the plastic DVD-ROM and sold you their Ultimate Gold-Plated Version at the very same price to themselves. Both versions of software, high end and low end, cost exactly the same to manufacture and distribute. In order to make a cheap version, they make the really good version, then break it on purpose.

They stick you with that barely workable, college-student, crippleware version so that they can engage in market segmentation. This allows them to suck money out of your pockets and a rich person's pockets at the same time. She gets her price, you get your price, and they make more money all round.

This practice isn't restricted to software. Much the same goes for hardware—for instance, laser printers. The fastest printer on the market was created with the best research-and-development effort. The slow printer was built with the same techniques and knowledge but has been deliberately screwed up. It has expensive extra hardware installed in it, to make it run slower. Why? Because if there wasn't a lousy printer on the market, then rich folks wouldn't pay extra for the best one.

Has it ever occurred to you that Federal Express might be slow on purpose? Most of the time, they could bring you that package at the break of dawn, at the very same cost and effort to themselves. But if they did that, you wouldn't pay any extra for their top-end service. So the drivers dawdle around until afternoon, just to make absolutely sure that you're not too gratified and well served.

These evil little aspects of commercial marketing are not cybernetic. They are perverse because they are about rela-tionships. They're cultural, they're psychological. Everybody in the restaurant biz knows that the most popular wine is the second cheapest. The second-cheapest wine is where restau-rants make all their money. They never sell much of the cheapest wine. They put it on their menu to make that second-cheapest wine look like a real bargain.

Remember extremeness aversion? It's an innate and irra-tional part of human psychology, hardwired into us through millions of years of quick, snap decisions. If we're offered three versions of a product—little, big, and middle—we go for the middle. It feels safest there. Therefore, a "medium" Coke at a theater can be about half a gallon of pop. As long as the jumbo version is a full gallon by contrast, we'll buy the half gallon and feel pretty good about it. The "small" is more cola than you can drink, but just try to force yourself to order a small, especially in front of your date. See how much that hurts?

There's nothing you can do about that feeling. You may get over it by a deliberate act of will, now that I've told you about the trick. But you'll still fall like a ton of bricks when confronted with three kinds of rental cars. It has nothing to do with the "information" that an "economy" car may well be the size of a 1940s limousine. In order for that information to affect your economic decision, you would have to stop to think about it. Thinking seriously about money is, as I pointed out earlier, a big stuffy drag. So you *won't* stop to think about it. Stopping your life to think about the details of revenue and expenditure is just too much of a hassle. The pantaloon, however, will do that sort of thing.

Because how much does money matter, really? You may not have paid for this book. Maybe you checked it out from a library, maybe it was a gift, maybe a friend lent it to you. I'm not all upset about that, even though I supposedly lost some money by it. I'm not going to blow a police whistle and fetch a shotgun. Because I may not require that money. For instance, someday I'll be dead. In fact, I may be dead right now. Hello there, reader! I'm dead! And I've got no revenue stream whatsoever! But you know what? I'm still talking anyhow!

So now you know about the information economy, how it wrenches money from people, and how its innovative craziness has been directly related to its profit margins.

Unfortunately, innovative craziness is a great way to lose money as well as make it. If you've got it made already, why take risks? If you bet double or nothing with a quarter of the market, you might get half the market. If you've got 90 percent of the market, like Microsoft does, you'd be nuts to risk all that. You want to sell new decals and tail fins, enough to move new models out of the showroom, but it's no longer much use to mess with the engine.

A perfect economy would be rational, but people are neither perfect nor rational. Their irrationality is the source of their exuberance and the ultimate source of wealth. Nobody gets out of bed in the morning thanks to cost-benefit analysis.

Historically, the upside dominates economies. People in a booming stock market foolishly think they can make money forever. People in a crash are much more chastened and "realistic," but they're just as foolish; they'll cut off their own nose for penance and short-sell their own children. If you're betting on the shape of future economics, expect booms. Booms tend to rumble on quite a while; busts last only three years or so before someone starts shooting.

The cheapest and most dependable way to make money in a market is to just buy the entire market. Really, don't try to second-guess anything; don't try to outsmart the rest of the human race. Just buy index funds, which require no concentration, no analytical genius, no transaction costs, and certainly no guru insights into the future from science fiction writers. An index fund is a simple bet on the persistence of business itself. Index funds do require a certain blind faith that there will always be some pantaloons around, but if there aren't, investments will be the least of your problems.

The other way to get hugely rich in the future is to spend incredible amounts of time, thought, and energy doing something really unstable, uncommon, and complicated. This is the Bill Gates method, the Steve Case approach, and the Warren Buffett routine. CEOs traditionally make far more money than even the most technically gifted programmer, scientist, or engineer; they can outearn lawyers and bankers. Unlike some passive rich idiot who's merely clipping stock coupons, an entrepreneur or chief executive officer can hire people, fire people, deploy resources, take up innovative opportunities, and, in some modest way, affect the general course of human events.

If you're a person of this kind, you probably already know it. You're antsy, you're blazingly determined and ambitious, you certainly don't need any futurist to tell you where you're going. To hell with making investments; that's no life for the likes of you. Pick up the football and carry it; *become* the investment. Make people give you their capital, make them invest in *you*.

Success is possible here, but it requires so much time and energy that money becomes a byproduct. Great success in business comes without guarantees, and the price is one's lifetime.

The information economy still has a lot of novelty-generating potential, despite the best efforts of all those who cashed in before you. There should be big money there for decades to come. However, as the new century dawns, the scene with real pits, pinnacles, booms, busts, and big, giddy air pockets is biotechnology. The range of gain and loss here is incalculable; that's not an information economy but a life economy. Its raw material isn't data, it's flesh. Computers and networks will continue to saturate the fabric of daily life, but DNA has just been mapped. Biotech is a different century's baby, and by its nature, it is much, much riskier and more potent than computers. If you go there, do keep in mind that although ones and zeros can bamboozle you, biotech can infect you.

The passion for wealth is a human passion. Like the other stages in the Seven Ages of Man, the pantaloon is a passing stage. This passion cools like the others do. The very rich are often at their most poignant and interesting in their endgame: the days when they deliberately and cool-headedly pick apart their own empires.

Only a true Scrooge, a genuine miser with a profound emotional deficit, finds any satisfaction in counting and recounting coins. Once you lose the passion that caused you to build a fortune, you can still find gratification in spreading wealth around. This doesn't mean relaxing (overachievers rarely enjoy much of that), but it does mean a different psychic landscape, a genuine change of pace. It means human engagement with the nature of time. It means wisdom.

A captain of industry who lacks a succession plan cannot call himself a professional. By making room for what comes next, he proves that he knew himself as well as he knew the business plan. Farsighted rather than merely on the make, he knew that his time, too, would come. That someday he, too, would have to turn the page.

MERE OBLIVION

> *Last scene of all,*
> *That ends this strange eventful history,*
> *Is second childishness and mere oblivion,*
> *Sans teeth, sans eyes, sans taste, sans everything.*

This book is about the future, but the clock ticks for it just as it does for every other human artifact. *Tomorrow Now* is necessarily a product of its time. It dates from an extensive belle époque, the historical period that began in 1989.

Historical periods do end, and they end in one of two ways. They jump or they get shoved.

The original belle époque underwent the shoving option: thanks to a terrorist provocation during a great-power rivalry, it crashed into the holocaust of World War I. The survivors then renamed the lost period Belle Époque because it looked so lovely in retrospect.

Or a historical period can jump; it can change through its own success. Feudal courts becomes national states, agrarian democracy becomes industrial. Failure may be crushing, but there is no transformation so unquestioned as a success. A successful historical period fulfills its own more-or-less manifest destiny, and more or less gets what it wants.

Which will it be for us?

September 2001 saw a determined effort to instigate another world war. If major land wars do somehow break out, with populations and governments crushed and capital cities on fire, then people will find themselves in a turbulent and violent time, and in a burst of harsh martial virtue the immediate past will be written off as a gilded age, all squalid and overindulged and pampered, and fatally innocent, and possibly somewhat effeminate. That happened the last time a belle époque collapsed, and the parallels are rather uncanny.

The parallels are far from destiny, though. One is inclined to predict that this belle époque will persist. A great war requires some great military powers, ready and eager to risk all. The New World Disorder lacks the muscle for conventional warfare. The New World Order has plenty of muscle but a problem finding foes to war with.

This isn't to say that a general collapse is impossible. Civilizations have fallen to bandits before, and even though nation-states now have profound military advantages over a rabble of nomads, civilization is vulnerable. In this chapter we should give full heed to our trembling anxieties. After all, this is Stage Seven: Mere Oblivion, the end of this book, and its primary topic is death. This is the place to give frank consideration to all the morbid options.

I was born in 1954, and during the first thirty-five years of my life there was always a ready opportunity to put a swift and total end to everything. Back in that previous historical period, the cold war, we liked to call that prospect "mutual assured destruction."

The generation of the recent belle époque, those happy folk who came to adulthood since the end of the cold war, are getting a rather better understanding of theological ferocity lately. Still, they often forget the truly sinister atmosphere that dominated from 1945 to 1989. Back in those turgid days, huge, well-organized governments were ready to respond to their political or economic issues by roasting everyone wholesale.

Atomic Armageddon was a vital aspect of the cold-war frame of mind. Armageddon served as a mental shortcut through many of the period's deeper problems; it was almost a kind of prayer. People *enjoyed* the everyday prospect of imminent mass death, the exciting sense that, at any random moment, everyone you knew and everything you loved could be blasted. Sudden and total flaming extermination gave people of the cold war a sense of identity and structure. It had genuine mythic resonance. This mythology supported many of the activities and attitudes of the cold war that seem the most peculiar and alien in retrospect. The space race, for instance. The widespread fascination with hallucinogens. The fierce joy people took in disposability: pop art, happenings, paper dresses, inflatable chairs.

All that atomic technology has been inherited; it has wended its way down the historical stream. Though their number has declined since their peak in 1986, there are still thousands of atomic bombs, cocked, polished, and ready to launch. Dark suspicions abound of suitcase versions to be detonated by small groups of cranks. But there is no Bomb. The myth evaporated, because the social attitudes that once supported heartfelt feelings of holy dread have changed. The Bomb has become very downmarket. It is no longer majestic in the planet's corridors of power; it is forced to skulk the littered, hand-me-down landscape of the New World Disorder.

The actual likelihood of people really and truly getting atomically bombed is much higher today than it was during the cold war. A bomb, or even half a dozen bombs, could go off almost at the drop of a hat. Ever since 1945, every five years or so, another nation has learned how to make these bombs, starting with the valiant superpowers and grinding down to grim, backward regimes like North Korea. But there's no corresponding allure left, no psychically fulfilling Götterdämmerung.

During and after the obliteration of Hiroshima and Nagasaki, a nuclear detonation was truly a gesture to conjure with. It conclusively solved the problem of ending World

War II. It did a great deal to freeze national boundaries in place afterward. In the twenty-first century, however, an atomic bomb solves nothing in particular. If you're a jittery, troubled nation like Pakistan, the possession of nuclear bombs makes you *more* likely to be invaded and dominated, not less. A bomb nowadays is probably best understood as Genocide in a Can.

A regional nuclear war—between India and Pakistan, say, or Iraq and Israel—would kill many millions of people. It may have happened by the time this book sees print. This would mark a major transition, replacing the easygoing belle époque with a new, grim era of self-righteous, Nuremberg-style retribution. But there would be no mutual assured destruction and no Final Solution. There would be just a very dirty, very gruesome complication, and when the smoke cleared, as the smoke would have to do eventually, the authors of the atrocity would probably be hanged by their own people.

The same applies to biological and chemical warfare, demigods of the same mythos with a cheaper operating system. Bugs and gases have always lacked the atom's mythical cachet because they are so cheap and easy to make. Small groups are willing to produce and use these weapons, as the Aum Shinri Kyo cult demonstrated when they scattered anthrax and also poisoned the subways of Tokyo with sarin gas.

Genocide, however, is not a coup d'etat. Aum Shinri Kyo had no chance whatsoever of governing Japan. Not only

were they insane; they simply lacked the manpower necessary to run a nation.

Biological and chemical attacks might very well usher in
a grim and mordant surveillance-security state, busily sniffing the contaminated air and wiping all surfaces with cotton
swabs. But how exactly are the perpetrators to profit from
that? Individuals can bow to a protection racket, for they
know the gangsters personally. But it's hard for an organized
state to bargain with a secret gang of surreptitious poison-
ers. This problem runs deeper than mere national pride
and honor: there are no verification methods. You can't tell
whom you're bribing. Invisible people can't guarantee an
end to the hostilities. A local mafia can and will ruthlessly
exterminate the rival criminals in its turf, but a gaseous
international conspiracy can't defend its own territory.
Lacking laws, codes, customs, and legal procedures, they
can't even rein in their own rogue elements. Secret conspira-
cies therefore make very bad treaty partners. They can commit great outrages, but they lack a way to consolidate their
gains.

Plagues by their nature are the clumsiest of weapons, for
they quickly spread to random parties. What is the victory
condition in a war of contagious diseases? Plague may or
may not be "the poor man's bomb," but there's no question
it's a poor man's doom. Areas with organized governments
and public health systems will be the *last* to collapse from
germs and viruses, not the first.

The worst wrinkle to mass-destruction schemes is the me-too factor. Grand Guignol horrors do have a great moral impact, but where is act 2? It takes truly fantastic moral arrogance to assume that those of your cause are the only people capable of using a dreadful weapon. If these kinds of "arms" become the weapons du jour, copycats are guaranteed. Successful blackmail guarantees more blackmail. New terrorist parties will arise quickly, similarly equipped with bugs and gases. They will almost certainly be strongly opposed to the aims of the first terrorist parties, for their grievances are just as sharp, and they don't want to be publicly shown up.

If a general chaotic slaughter starts, it'll end up where the planet's forces of ruin are always most acute—inside the disorder. Plague would not be of much use in blackmailing rich, healthy nations. It makes more sense to use plague to ruin vulnerable poor nations and to render them ungovernable. A well-earned reputation for permanent bad health would create a useful ghastly pall. The pestilences would keep busybody global police at bay, along with annoying charity workers and globalizing tourists. Then one could grow illegal drugs in these septic badlands, exporting *that* plague to eager consumers worldwide. That approach might pay.

Some theorists of "asymmetric warfare" argue that, as human knowledge inexorably advances, smaller and smaller groups will surely obtain and use larger and larger weapons. We'll logically reach the point where one brainy guy in his basement can end the whole world. And this is our doom. Clever enough to torch everything on a whim, we humans never quite got it about creating fire departments.

There's some plausibility to this line of thought, in the dark conviction that our species will outsmart itself to death, because human knowledge is simply not compatible with human survival. Atomic physics hasn't done that trick for us yet, but surely any line of scientific inquiry, pursued far enough and deep enough, would yield some handy method of massive destruction. It is an existential dilemma.

If that's really our predicament, we might as well honestly face up to the very dark implications. Okay, fine; it may well be that small, demented cults armed with cheap yet catastrophic technologies will cheerily bring a horrid ruin to all and sundry, including themselves. We humans will leap downhill into a new dark age from sheer spite and our innate dementia. Should that happen, then we humans pretty clearly have that fate coming. We will deserve it. It means that we're just not up to the difficult task of being aware, intelligent, and civilized. After a couple of geological epochs, maybe the raccoons will take the stage and do rather better than we did.

Now we'll put that prospect aside. One should never get all glumly fixated on the latest and sexiest version of apocalypse. It is self-indulgent laziness to predict that the clock will stop ticking—just because it happens to be our watch. Nothing looks sillier in retrospect than an antisocial prophet ranting that a great flood is due to cleanse the world of other people's wickedness.

When it comes to Mere Oblivion, we've not yet scratched the black and adamantine surface. Human wickedness may indeed be a fearsome matter, but we shouldn't flatter ourselves. There are far more threatening forces in this world than our lust for self-immolation.

For a truer understanding of the fragility and contingency of life, we need to lift our heads. We need to look to genuinely cosmic disasters.

This requires some mind stretching, but it's important to grasp the metaphysical fact that the cosmos does not owe humankind a living. Seen on the proper scale, our planet is debris.

It doesn't seem likely that our Sun will suddenly emit a giant flare that cooks Earth, but if it does, we have no useful response. We can't do one single thing about the Sun's misbehavior. We're similarly helpless against all the other local stars as well. If a supernova occurs anywhere in our galactic neighborhood, we will be flooded with searing cosmic rays, in which case we are doomed.

As far as we can tell, that hasn't happened in our galactic neighborhood in the past four billion years, but it certainly happens. If stars didn't blow up, then Earth wouldn't be here. Every earthly element heavier than iron was created in the bowels of exploding stars. We're made from detonated star junk.

Furthermore, our local region of outer space is rich in giant, flying mountains of ice and rock, many of them untracked and unnamed. The belle époque year of 1994 offered a striking hint when the Shoemaker-Levy comet very loudly and publicly fell onto Jupiter. That comet broke into twenty-one separate chunks, and if even one of those mile-wide chunks had hit Earth instead of Jupiter, it would have left a crater bigger around than Washington, D.C., and deeper than twenty Washington Monuments stacked on end. Had that happened, you would definitely not be reading this. Any people still alive would be shivering under the blackened, freezing, acid-bitten skies of an "impact winter."

We have Jupiter's experience on tape. It's also in the fossil record. In its long and checkered history, life on Earth has undergone five truly major-league calamities. A disaster 438 million years ago killed off 85 percent of everything growing, breathing, or moving around. The Devonian period met a similar brutal end 367 million years ago, when 82 percent of all species were lost. The third and greatest catastrophe occurred 250 million years ago, at the end of the Permian period. This amazingly dreadful event killed off as much as 96 percent of earthly life and, incredibly, almost managed to exterminate the cockroaches. The Triassic period ended in a fourth horrid orgy of destruction 208 million years ago, killing off three-quarters of everything. Great calamity number five, the famous Cretaceous-Tertiary boundary event, hogs most of the publicity, because it caused the death of the extremely photogenic dinosaurs. It resulted in the loss of a respectable 76 percent of all species.

It's pretty clear now that the Cretaceous-Tertiary boundary event was caused by a large meteor hitting the area we now call Mexico. Similar suspicion is building about the other five events, although they might have been caused by solar climate change or by giant volcanoes. The depths of Earth and the core of the Sun aren't our friends, any more than the comets are.

These may seem like remote happenings. But though mass extinctions of life on Earth don't happen every Tuesday, they happen. We are currently experiencing the planet Earth's sixth major extinction event. Life here is undergoing a bizarre calamity: a searing, all-consuming outburst of intelligence. Tens of thousands of species are perishing.

Sucked into a whirlwind of global transformation, the planet has become a biological melting pot, where ancient species native to profoundly different environments are suddenly at war, everywhere. North America in particular is a vast patchwork of imported alien species. Environmental evildoers such as kudzu, zebra mussels, fire ants, and witchgrass are just the best-publicized offenders.

Humans get a lot of press when it comes to environmental ruin, because humans love to talk about themselves, but many other species, pursuing their own selfish interests, are also major environmental offenders. Unlike humans, these highly competitive entities have no interest in ethics, ecology, or legislation; they are remorseless and very powerful.

As we learned in stage one, every human body is a bacterial ecosystem of many hundreds of species. These species include not just our brimming bonanza of microbes and viruses but human-specific fleas, lice, and eyelash mites. These species go wherever we go, and so do burrs, and seeds, and sprigs, and flowers, and rats, and dogs, and crops, and all our beloved favorites. A growing tonnage of human flesh and human symbionts commands more and more of the planet's biological fertility. For the species elbowed aside, extinction is a necessary consequence.

Then there are weeds and animal invaders. They are a very serious matter. Even if human beings vanished tomorrow, the extinctions our intelligence provoked would continue for millennia. Because even though we humans rather like wild creatures and often take remarkable steps to protect them, we can't effectively protect plants and animals from one another. Imported fire ants are slaughtering native species in my front yard right now, but there's little effective recourse, short of fantastically comprehensive acts of fire-ant extermination across thousands of square miles of the southern U.S.A. Goats and cats, released onto small islands, are a major, serious menace to many endangered species. The inhumanity of animal to animal deserves more respect than it gets.

It's hard to recognize that a neatly groomed lawn with a little kid, a puppy, and a kitty is a biological holocaust. But it is. Whenever you witness a lovely sight like that, it means that half an acre of the planet's surface, which formerly supported many hundreds of various weeds and beetles, has been reduced to just four species (not counting their microbial inhabitants). That is the true face of the Sixth Great Extinction. It's a face that we humans find pleasant. It's not a mean guy with a club skinning a seal. It's civilized people playing badminton on the lawn, maybe having a lemonade.

This stark realization can cause despair, but wait—there's so much, much more! Terrific ecological damage was done to the planet 250,000 to 10,000 years ago, when the planet's intelligence catastrophe was just ramping up. When modern humans first arrived on the planetary scene, Earth was in the midst of an ice age, with mile-high sheets of ice depressing the continents. By Earth's normal climate standards, pickings were pretty meager. Nevertheless, this pinched and chilly world was roamed by colossal living hordes of meat on the hoof: mammoths, wooly rhinos, giant bison, plus the very large ferocious animals that preyed on them, such as dire wolves and saber-toothed cats. Today there are still a few Asian and African herbivores that weigh over a ton. All the rest of the really big meaty animals disappeared millennia ago.

Besides the jumbo-variety prey of the thousand-kilogram-plus variety, there were plenty of species weighing one hundred to one thousand kilograms. After modern human beings arrived in the vicinity, three-quarters of them vanished. Of the many tasty animals from five kilos to one hundred kilos, about half went extinct. Then there were a lot of runty Pleistocene critters running around that no sane person could be bothered to eat. Only 2 percent of those creatures vanished. This sixth mass extinction has been a very choosy one. And it's been going on for quite a long time.

So let's be entirely clear: human beings *are* the Sixth Great Extinction. We humans have never lived in any state of balance with nature. There were quite a few ancient species of prehumans and parahumans that more or less managed this feat, but once we showed up, they vanished just like the ground sloth, and probably for the same reason.

Until quite recently, we humans had no idea what we'd done. We certainly had no moral choice in the matter. It was just one of those things—the sixth one of those things. We were clever enough to wreak havoc but not clever enough to measure it, record it, study the consequences, and take any preventative measures. The scale of the trouble was beyond our mental grasp.

This worldview is different from any we've had before, and it comes with important implications. For instance, all of humanity's modern technoindustrial wrinkles, such as superhighways, mass production, nuclear plants, almost every unnatural gizmo decried by the many human friends of nature, in other words—all these are merely a topper on our 250,000-year spree. We severely disrupted nature long before we invented written history. Until very recently, we modern humans have had no idea what a truly natural world should even look like. Because a natural world looks pre-human. No human being since long before the birth of agriculture has ever witnessed a state of nature. There are no writings about it, no photographs, no records or documentation. The only portraits of it are on cave walls.

For my own locale here in Texas, the state of nature entails giant armadillos, sloths as big as hippos, three kinds of elephants, carnivorous long-legged pigs that can run like antelopes, condors with wings sixteen feet across, plus llamas, camels, giant bison, giant wolves, and giant bears. Minus human beings, those animals would all still be here. A natural Texas would look like the Serengeti on steroids.

So when we Americans are marveling at our moose, elk, the remaining bison, and so on, we're witnessing a radically impoverished fauna, the shrimpy leftovers of an ecological catastrophe. Even animals we Americans consider symbols of wilderness have been here just five hundred years. In Texas our much-cherished mustang horses and longhorn cattle, even our tumbleweeds, are imports from Europe.

I know this worldview sounds rather far-fetched, as if I'm waxing science-fictional for the sake of mere paradoxical exoticism. But the truth is often fantastic. If science fiction has any truly profound insight to offer us, it's that existence really is weird. Human ideas of "normality" are always merely local and temporal. As in Einsteinian relativity, everything must change with the chosen framing device. It's not the end of the world to realize that existence is weird; it's not the end of anything. It's a matter of perspective.

In the long perspective of the species Homo sapiens, the boring everyday world that we inhabit is very far from typical. People were walking around for 240,000 years before they ever started getting fussy about way-out notions like literacy and irrigation. If you look at us from the perpective of our geological time span rather than through our written history, then it's obvious that a normal human life is prehistoric. The default state for humans is a clan of preliterate hunter-gatherers. The statistically normal experience is squatting in the tall grass with a dozen of your relatives, watching large tasty animals through slitted eyelids.

People who write history have a natural contempt for the illiterate, so they gloss over everything that lacks documentation. Prehistoric people weren't as hapless and goofy as we historic folk like to make them look. For instance, contemporary images of cavemen carry clubs, but no actual caveman anywhere has ever been found with a club. Furthermore, most prehistoric people never had much to do with caves. Caves are not particularly attractive places for human beings to live; they just happen to be very good places in which to find prehistoric evidence. Our notion of a caveman always wears ridiculous clothes, too: that ragged bearskin rug with a neck hole. Two-million-year-old pre-human australopithecines may conceivably have dressed that badly, but human beings are rarely so unstylish.

For a better idea of the long-lost Stone Age, we need to move a hero front and forward, a kind of Everyman figure for Stage Seven: Mere Oblivion. So let's consider Otzi. Otzi makes a wonderful protagonist for a chapter on death, because he is one of the world's longest-dead dead guys. He is also the best preserved. Otzi is a Stone Age gentleman who was revealed to us in the belle époque year of 1991 by the thawing of a European glacier.

The most striking thing about Otzi is his nifty kit. He is prehistoric and preliterate, but he has a little chin-strap hat, a knee-length woven-grass raincoat, and a reversible vest with stylish black-and-white stripes. He also sports leather shoes and a wide variety of ingenious knickknacks: an axe, a bird-catching net, a fire box of birch bark, a flint scraper, a horn drill, a stone awl, and about a dozen other keen tools and toys. He has everyday personal uses for sixteen different kinds of wood. He even has tattoos. Otzi lived and died around 3300 B.C., but that doesn't mean he was meandering around "being primitive" for our benefit. Otzi was very clearly an everyday guy with a lot of regular, sensible routines: tying his shoes, stitching up his hunting coat, and chewing some kind of gum (it left telltale spots on his teeth).

The only thing truly out of the ordinary about Otzi was his unhappy end. Due to foul play, while mountain climbing in the Alps, he lost or broke his bow and arrows, and also he broke his ribs. So Otzi ended up getting trapped by a snow-fall and preserved in a mountain glacier, so that people two hundred generations later can marvel at him and consider him some kind of freak. Otzi suffered an unusual and fatal mishap, but statistically speaking, he is the single most normal and average human being we can see in the modern world.

This Stone Age man has four pieces of stone on him (and one very exciting piece of futuristic copper metal), but almost all of his kit is totally biodegradable. If he'd been buried properly and dug up nowadays, we'd have known nothing at all about his tailored pants, his raincoat, his belt, his back-pack, and especially a couple of dozen different kinds of thongs. To judge by Otzi, the Stone Age was really the Thong Age: the guy never made a move without a handy reservoir of string.

Okay, granted: at a mere 3300 B.C., our friend Otzi was late Stone Age, not deep Stone Age. He was a contemporary of the early Sumerians, and he came from a Neolithic village where they grew grain and had some livestock. But to under-stand how the rest of the world has suffered from the Sixth Great Extinction, you have to imagine a natural world in-fested for 240 millennia not by dumb, sluggish cavemen but by guys like Otzi. It's a world where creatures twelve feet high at the shoulder were obliterated by encounters with humans. Incredible, unnatural predators turned loose upon the unsuspecting Earth. Intelligent predators who can chase whole herds over cliffs with wildfires, poison them with herbal venom, dig huge death traps in the dirt. Creatures able to disguise themselves with their victim's own skin and then mortally wound them from a hundred feet away. Intelligent omnivores that could skin and swallow almost anything. No other predator on Earth can do that. No other predator can even *imagine* doing that.

Otzi came out of a melting glacier. It's no coincidence at all that he showed up among us at this point in time. Glaciers are melting all over the world, just as glaciers melted during the Stone Age, only on an industrial scale—much, much faster. Here is a little list of melting glaciers, circa the year 2000, courtesy of Worldwatch Institute, which makes it its business to keep an eye on this sort of thing.

Greenland Ice Sheet—Greenland
Has thinned by more than a meter a year on its southern and eastern edges since 1993.

Columbia Glacier—Alaska, United States
Has retreated nearly 13 kilometers since 1982.

Glacier National Park—
Rocky Mountains, United States
Since 1850 the number of glaciers has dropped from 150 to fewer than 50.

Tasman Glacier—New Zealand
Terminus has retreated 3 kilometers since 1971, and main front has retreated 1.5 kilometers since 1982.

Meren, Carstenz, and Northwall Firn Glaciers—Irian Jaya, Indonesia
Glaciers shrank by some 84 percent between 1936 and 1995. Meren Glacier is now close to disappearing altogether.

Dokriani Bamak Glacier—Himalayas, India
Retreated a total of 805 meters since 1990.

Duosuogang Peak—Ulan Ula Mountains, China
Glaciers have shrunk by some 60 percent since the early 1970s.

Tien Shan Mountains—Central Asia
Twenty-two percent of glacial ice volume has disappeared in the past 40 years.

Caucasus Mountains—Russia
Glacial volume has declined by 50 percent in the past century.

Alps—Western Europe
Glacial area has shrunk by 35 to 40 percent, and vol-

ume has declined by more than 50 percent since 1850. (It's thanks to this phenomenon that Otzi suddenly showed up.)

Mount Kenya—Kenya
Largest glacier has lost 92 percent of its mass since the late 1800s.

Speke Glacier—Uganda
Retreated by more than 150 meters between 1977 and 1990.

Upsala Glacier—Argentina
Has retreated 60 meters a year on average over the last 60 years.

Quelccaya Glacier—Andes, Peru
Rate of retreat increased to 30 meters a year in the 1990s.

This is due to a global phenomenon called the greenhouse effect. The greenhouse effect has an unfortunate name, because a greenhouse is a nice, warm place where kindly gardeners shelter rare and valuable plants. In the twenty-first century the greenhouse effect will likely get an exciting new name, perhaps something like "weather violence" or "ecogenocide."

The greenhouse effect is the dirty little sister of nuclear Armageddon. They're both about technological energy sources. However, while a nuclear bomb lets a great deal of energy loose all at once, fossil fuels slowly spread their pall of wasted energy throughout the atmosphere of the whole planet. The menacing greenhouse wears the Bomb's hand-me-downs.

Throughout my lifetime, carbon dioxide has increased in the atmosphere at one part per million per year. It stood at about 310 parts per million when I was born, and it's now at 360 and trending up. What's the problem with this? Well, a carbon dioxide molecule flying around in our atmosphere absorbs infrared radiation. Specifically, it absorbs it at 2.5-, 4-, and 15-micrometer wavelengths. That means that infrared energy that once would have fled into deep space is instead captured by carbon-dioxide molecules and kept inside the planet's atmosphere.

Carbon dioxide isn't the only greenhouse gas, though it's the most voluminous. There are also methane and CFCs to fret over. Put together, all these waste gases in the atmosphere are adding about two and a half watts of captured energy for every square meter on the planet's surface. This sounds like a very abstract problem, what with the wavelengths and the statistics and all, but there are ways to imagine it that can get us a little closer to the fire. Two and a half watts is the energy in a burning birthday candle. A square meter is about the size of a tabletop. So just imagine our whole world paved in tabletops, from pole to pole, and every single one of those tabletops has a little birthday candle on it, burning constantly. That's what we've done to ourselves with the greenhouse effect. That's why those glaciers are melting.

The greenhouse effect comes from digging up fossils (fossils much, much older than Otzi) and setting fire to them. We humans have been doing this seriously for two hundred years. Setting fire to long-extinct life-forms is the human race's primary industrial enterprise. Everybody is implicated. This thriving business underwrites almost everything we do.

Coal and oil and gas are gigantic, highly profitable industries. Almost everybody on Earth is dependent on fossil energy. Even remote nomads in the African Sahel have radios, plus airborne supplies of tractor-grown charity food when their climate misbehaves. It's the Pleistocene situation all over again, but with thousands of times more humans.

Unlike the mythos of the Bomb, this isn't about human extinction. The massive die-off in the Pleistocene did not kill off humanity. But it did incredible numbers of just about everything else. Habitat disruption and species replacement are mighty engines of mass extinction. A full-blown greenhouse effect guarantees both.

There are a lot of humans around now, about a hundred million tons of living human flesh. We humans outweigh all the wild mammals in the world ten times over. Anything crushing us will crush wild animals first and worst. So it's nature that is going to catch it from the greenhouse effect, not people.

For humans, the most dangerous aspect of the greenhouse effect is the prospect of crop failures. Crop failures could easily starve many hundreds of millions of us. However, crop failure won't kill people in the *industrialized* world, the people spewing most of the carbon dioxide. The industrialized world has sophisticated transport and storage systems, powered by fossil fuel. That almost guarantees food. It's people living close to the soil, under nearly natural, subsistence conditions, who are in great peril from the greenhouse effect. What little order they have, they will lose.

Our friend Otzi was killed by a change in the weather. Otzi didn't intend to freeze to death in a sudden, unseasonable storm. Otzi was a serious, businesslike man, pushing forty, the kind of guy who has an agenda, probably a wife and kids. But something, or somebody, broke his ribs—and to further his discomfort, he had a small flint arrowhead stuck in his back.

Some people think he would have died of that stab in the back, but despite the pain of his broken bones and that chip of flint in his flesh, he was tough. He was composed enough to carry his possessions, methodically create some new ones, and even sit and eat a final meal.

Otzi was heading for a mountain pass that he knew well. He was making a tactical retreat of some kind. He was arming himself with a new set of freshly carved arrows. Had he succeeded in his plans, the people who shot him would very likely have lived to regret that. As an experienced alpine hunter, Otzi must have known there was some risk from the weather. But in the stress of his circumstances, he weighed his risks incorrectly. So bad weather caught him, and bad weather did him in.

P eople have been talking about the greenhouse effect for more than a hundred years. The trend is now very clear. If the twenty-first century cannot find some better way to power itself, it means ruin. That doesn't mean that we'll all perish like Otzi. Most people are too smart and well organized to get killed in large numbers in "natural" calamities such as greenhouse droughts and giant greenhouse floods. We don't find whole frozen armies of Otzis. Bad weather generally picks humans off, here and there, in ones and twos.

But the effect on wildlife and the general environment will be devastating. Over a period of decades, shifting weather patterns, consequent extinctions, and the sheer airborne filth that comes from burning fossils to stay alive will slowly degrade and defile everything we know. It will ruin all the beauty of our belle époque, transforming the whole Earth into something like a grim mining town in East Germany, only without frogs.

A world that runs on smoke is a chronic problem, like smoking cigarettes. There is no particular cigarette that kills you. It's not a sudden deadly calamity to rip open a pack and fire one up. But smoking pack after pack, year after year, means ruin. Not in any immediate way. Ask for "scientific proof" that a cigarette has given you emphysema or cancer. You'll never get any. That is a loophole that the tobacco industry rolled through for decades on end, just as fossil-fuel industries exploit it now. There can be no such proof. There is merely a generation of tobacco users who are wrinkled and wheeze, while some have deadly illnesses.

The only evidence of harm you get from cigarettes is "anecdotal," such as that bad taste in your mouth, and your yellowed teeth, and the sore throats and persistent colds, and then the coughing and the shortness of breath, and the odd tightness around your heart, and so on. All these phenomena have alternate explanations. If you are stuck on your addiction and unable to face reality, then they can all be explained away. Unfortunately, they never really and truly go away until you quit smoking cigarettes.

Here are a few greenhouse symptoms. These are typical ways in which the planet manifests this trend.

Eastern U.S.A.
July heat wave, 1999. More than 250 people died as a result of a heat wave that gripped much of the eastern two-thirds of the country. Heat indices of over 100 de-

grees Fahrenheit were common across the southern and central plains, reaching a record 119 degrees Fahrenheit in Chicago.

Texas

Record downpours, 1998. Severe flooding in southeast Texas from two heavy rainstorms with 10- to 20-inch rainfall totals caused $1 billion in damage and 31 deaths.

Florida

Worst wildfires in 50 years, 1998. Fires burned 485,000 acres and destroyed more than 300 homes and structures.

Mediterranean

Intense drought and fires, 1990s. Spain lost more than 1.2 million acres of forest to wildfires in 1994, and 370,000 acres each burned in Greece and Italy in 1998.

Indonesia

Burning rain forest, 1998. Fires burned up to 2 million acres of land.

Eastern U.S.A.

Driest growing season on record, 1999. The period from April–July 1999 was the driest in 105 years of record keeping in New Jersey, Delaware, Maryland, and

Rhode Island. Agricultural disaster areas were declared in 15 states, with losses in West Virginia alone expected to exceed $80 million.

Western U.S.A.

Year 2000 was the worst fire season in the United States in 50 years. Eighty-four large fires burning in Arizona, California, Florida, Idaho, Montana, Nevada, New Mexico, North Dakota, Oregon, South Dakota, Texas, Utah, Washington, and Wyoming. Federal troops were called out to fight fires.

Mexico

Worst fire season ever, 1998. 1.25 million acres burned during a severe drought. Smoke reaching Texas triggered a statewide health alert.

This last symptom—the 1998 Mexican event—marked the moment that made me a greenhouse believer. The sky above my hometown was the color of television for about two weeks. I'd never seen any weather event so uncanny. Later, in the summer of 2000, it was 112 degrees in my front yard, the highest temperature ever recorded here.

I should stop reciting symptoms now, because I'm finding this a little embarrassing. As a futurist, I know that people in the next few decades will be looking back on a list of contemporary horrors with an indulgent and dismissive smile. "You thought *that* was the greenhouse effect? Ha!" They'll be doing things much more hands-on and visceral: stacking sandbags, digging wells.

In 2000 nobody voted to drown Mozambique in three major tempests. But if a cigarette is a "coffin nail," then a fossil-fueled light switch and an ignition key are your personal hands-on interfaces to a planetary gas chamber. If atomic Armageddon was like watching your neighbor wire a shotgun to your forehead, then the greenhouse effect is like putting a cigarette in your own mouth. There's no doubt that the greenhouse is here now, and that worse is on the way. The only doubts are about how bad it will get and how long it will take us to effectively address the problem.

We have two basic choices: jump or get shoved. We can jump into a new world of cleaner energies, or we can get pushed there, probably at gunpoint.

With this dark, enormous specter literally filling our sky—tanker trucks full of it, traffic jams smoking vigorously, oil wells on fire as camou helicopters hover over the spreading disorder—well, it's high time to consider a kindlier scenario. Suppose that we jump.

All things must pass, but the end of an era is not a synonym for its defeat. If a student becomes a lover, or a warrior a judge, that's cause for celebration. What happens to our belle époque, when and if it gets what it really wants?

Getting what you want is a serious matter. It is far more transformative than frustration. Unlike defeat, which can pass, success is guaranteed to change you, giving you radically new circumstances and a whole new set of motives.

If there's a single word that sums up the secret hopes and the sweaty, fervent desires of the belle époque, it is "posthuman." The posthuman is the dreamy vision that the belle époque clutches to its panting bosom. The posthuman is the belle époque's own heartfelt version of the technological sublime.

What is "the technological sublime," and why do we swoon for it? Why do we fall for this silly-sounding thing? It is the projection of a spiritual need for transcendance onto mechanical hardware. A sublime thing inspires awe and wonder. It ruptures the everyday. The sublime is a liberating spectacle that lifts the human spirit to plateaus of high imagination. The most alluring and attractive way to deal with technology is to hype it as something divorced from the normal routines of life.

The roots of the technological sublime can be traced to twelfth-century Europe and a sudden competitive outburst of high-performing, high-tech Gothic cathedrals. Since then, Western civilization has lacquered spiritual transcendence onto technological innovation whenever it gets the chance.

Print becomes Gospel. Electricity is the Life Force, and so is oxygen, or phlogiston. Steam is Power itself. Telegraphs abolish Time. Railroads abolish Space. Aircraft become Wings Across the World. Cars become On the Road. Hallucinogens become Enlightenment. Uranium fission becomes Atomic Armageddon. Rockets Conquer the Stars. It's the standard method by which Western society promotes the changes it calls progress.

The technological sublime is not merely an illusory need to romanticize machines. Any genuinely new technology *really does* have some fantastic, transcendental possibilities. Flying, for instance, is unquestionably an astounding, world-transforming achievement. The process that necessarily follows—moving from gold medals and massive fame for Charles Lindbergh in 1927 to bored, jet-lagged, yawning passengers at the Orly Airport in Paris—is not disenchantment or disillusion. It is domestication, and that is the hallmark of success.

Sublimity is like the sense of wonder. It is an authentic human emotion, but the human sensorium cannot bear it for long. If everything around us were permanently sublime, we would be frozen in place with our jaws dropped, unable to think, unable to eat, unable to live. A person unable to wonder is spiritually dead, but this emotion is best directed toward phenomena all out of human scale, such as geological time spans or the physical scale of galaxies. If you're poleaxed with sublimity by some technogizmo made up by the guy next door, you'll gain nothing useful; you'll only have your pockets emptied.

The belle époque, like most epochs in the Western world, adores technology, but since the population in general has become technologically sophisticated, we're rather more blasé about it than we used to be. With the new notion of the posthuman, we've finally brought our gadgetry to bear on the core of the problem: ourselves. Instead of wishfully hoping that some machine will become our wings, we aim to transform humanity directly.

This is not a pipe dream. This is achievable. There is an almost infinite number of possible technical ways to transform humanity. We can start from our firm kinship with the microscopic and work upward through every scale. Genetic ways. Mitochondrial ways. Tissue. Bone. Nerves. Guts. Through blood, lymph, and hormone. Through our senses, through our neurons. We are large, physical, multicellular entities; every aspect of our being offers up a scientific, technical, and industrial carnival.

Almost anything that can be done to a lab rat can be done to a human being. Lab rats are presently undergoing extensive technical modifications—everything from spinal implants to genetic knockouts. It's no accident that animal-rights activists carry out their cultural warfare on this important front. The postrodent is the standard-bearer for the posthuman; if it's icky and unnatural on a nude athymic mouse, it'll look and feel a hundred times as icky on your grandchildren. Before posthumans will come postrodents, postfruitflies, postnematodes, and, very interestingly, post-human samples of formerly human tissue.

Fetal cell cultures, cancer cultures, human flesh in a petri dish, deprived of rights, citizenship, a budget or a vote, the ability to feel pain or register happiness or discontent—this parahuman material is of critical importance. It is the indus-trial resource from which all else posthuman will follow. It's the equivalent of crude oil for the plastics industry. Human cell cultures are a futurist consommé of tomorrow's human condition.

Biotech research is not "health research." To think this is a category error. Medicine is just one subset of biotech. Doc-tors are a social caste, not the masters of life and death. Biotechnology is not about restoring sick human flesh to baseline healthy performance. Biotech is about cultivating the ability to move the flesh of *any* species into *arbitrary* levels of performance. Therefore: grains that grow in salt marshes; goats that excrete hormones; mice that grow human ears on their backs; rabbits that glow in the dark.

There's a taboo about discussing these transformations openly and freely in relation to human beings, but that taboo is melting away year by year. A taboo is a very mild and weak barrier to technological change, compared with the very stiff barrier of not being able to do it. Once you're able to do it, then you can become persuasive. You can go for the enchantment of the technosublime, rephrase the arguments, and redefine the paradigm. You can make the taboo corny, another shibboleth of a dead epoch. Even the most fiercely held moral beliefs can and do yield to this treatment. "Racial science" meant everything to Nazis; "miscegenation" once made Americans panic. Viagra is considered a commodity, not an aphrodisiac. RU-486 is a morning-after pill, not an abortifacient. Law and philosophy can't trump engineering. In a world fully competent to command its material basis, ideology is weak.

The destruction of the human condition is not the fault of some sinister cadre of mad-scientist masterminds. Mad cadres of masterminds lack staying power. They are not serious players in real social transformation. The human condition is changing because this is what our culture genuinely *wants.* This profound desire can be seen in the public reclassification of degenerative syndromes into treatable illnesses. Senility becomes Alzheimer's disease. Menopause becomes something you take pills for. Wrinkles submit to Retin-A, and baldness to Rogaine.

Wrestlers become icons, not because wrestlers are hugely entertaining but because wrestlers are huge, unnaturally shaped human beings who can snort steam in public—and that is popular.

The female bodybuilders of the belle époque are women without historical precedent. Even genuinely ferocious and bloodthirsty women, say, the wives and sisters of the Mongol hordes, would flee in superstitious terror at the sight of modern female bodybuilders.

Modern actors and actresses, supermodels both male and female, have backs and stomachs ridged with muscle. When people like these are decried as "unrealistic role models," it's because they are role models for the *next* historical epoch, not this one. They're unrealistic, but only for *us*. When our version of "realism" is dead and gone, then, by definition, our epoch will have been been replaced.

We don't want realism; we want what we want. Nobody much wants to be Frankenstein or RoboCop. Those are mythological versions of a posthuman being. We might be forced into such molds by terror or state coercion, but that's not our desire. Our desire is openly displayed on every billboard, in every magazine, on every screen: everybody wants to be strong and attractive, indefinitely. It's Death, Mere Oblivion, as a treatable condition.

Humans are transformable; but how and where will this transformative power be deployed? That depends on the character of the society creating and sustaining this power. Belle époque society is not a chilly, abstract realm of Platonic philosopher kings. Nor is it a cold-war security state. The belle époque is a buzzing, heaving, global capitalist marketplace. Its most powerful and important institutions are not armies, states, academies, or churches. They are the World Trade Organization, the International Monetary Fund, and a motley host of highly wired nongovernmental entities, in and out of national boundaries, and in and out of the private sector. Standards boards; transnational infrastructures; quasi-autonomous nongovernmental organizations such as Doctors Without Borders, the European Society of Biomechanics, the Federation of Asian Pharmaceutical Associations. If the belle époque gets what it wants, then the posthuman imperative will follow widespread market demand, not any ethical agenda imposed by local boffins or pundits.

People will be given what they want. Tomorrow's end users/consumers will not be given what their doctors or their pastors think is good for them, for their bodies and their souls. Instead, people will be gratified. When this happens, the belle époque dies, because to get what you want means the end of what you were.

The posthuman doesn't mean Utopia (a synonym for Oblivion). Nothing is perfected, nothing is resolved. But it does means a new civilization with genuinely novel means of behavior and support. It is not a merely revolutionary change. It is a deep and permanent break in cultural and historical continuity. A revolution is just "the violent overthrow of one class by another." The "posthuman" necessarily means a redefinition of what it means to be alive. This is a crushing blow to many eternal human verities, among them death. One of posthumanity's necessary consequences will be the violent abolition of Shakespeare's seven ages of man. In a truly posthuman milieu, those terms of Shakespeare's no longer make sense. The natural processes of growth, maturity, and aging are altered, scrambled, shuffled, or abolished. Human life loses its natural arc. The human stage is cleared, and new players tread the boards.

What does this mean? How does it feel? It means that certain traditional phantoms of twentieth-century futurism are unlikely to happen. As we pointed out in stage one, there is very little market demand for a genetically altered human baby. Nobody with a spark of common sense wants to be the first customer to try this product, for as with computer software, the early adapter always gets burned.

Old people, however, are another arena entirely. There is well-nigh unlimited market potential in combating aging and death, especially in a society, like ours, with unprecedented numbers of well-to-do people past the so-called retirement age. There are legions of hardened street fighters willing to battle in clinics over the fate of unborn babies—but there are very few champions of death rights who will leap up and insist that they personally should be allowed to die. In any case, their success would quickly remove them from the debate. The customers for life extension will be very focused and committed people. Life is a seller's market.

When you pay for extended life, it will probably be much like paying for information. You'll never get the commodity itself, pure, immediate, and free. It will come in bits and pieces, full of qualifiers, caveats, snags, and trapdoors.

If the belle époque becomes the parent of the near-term posthuman, there won't be one large, dramatic break-through, such as a grand, government-sponsored moon landing or a military atomic blast. There will be dozens of medium and small advances. They will be webbed together in sophisticated ways, exploited by small, mostly profit-driven groups, working together electronically, over increasingly leaky and porous national boundaries. It won't look like Ponce de León's Fountain of Youth. It will look like the software industry, but medical. Juicier. Vastly more painful. Shipping behind schedule, wondrously botched, out of the doors of scanners, from the eyes of needles.

Paternalistic governments will not take stern steps to spread these benefits evenly, as they did with the twentieth century's universal suffrage or national phone service. They will be picked up first by technologically adept early adapters, then released in mass-market versions subject to periodic upgrades.

Most of the people working in the new industries will not be professionally trained. They will lack ethical codes and esprit de corps. They will be sucked in by market demand from other industries and disciplines. They will train themselves by studying the Net.

Successful companies will commonly disintegrate as everyone leaves for a start-up. Quieter areas of the industry, the parts that have dependable demographics and sustained demand, will be snapped up by huge enterprises seeking economies of scale. Certain areas of research and development will be palsied or killed by social resistance, transformed into smelly mires of cultural warfare and endless litigation. These prunable parts are the ones that can be zeroed in on by one sound bite, such as "Chernobyl," "Brent Spar," "Love Canal," and "Bhopal."

"Posthuman" is a sound bite. Like the term "cyberspace," it is a Procrustean neologism, raggedly covering some very ill-defined and multiplicitous ground. People who are genuinely posthuman will not consider themselves to be "post-" anything. The posthuman means an end to us and our concerns, but it's just a beginning for them and theirs.

Contemporary people find this situation very hard to imagine. In science fiction this problem of imagination is known as a "Vingean Singularity." A Singularity is a place where matters that would be of great importance and interest to futurists become impossible to describe, simply because we futurists are, in point of fact, human beings. Being human, we have inherent human constraints: cultural, verbal, cognitive, and so on. We have never been two hundred years old, we do not have an IQ of 312, we have no industrial-strength DNA inside our cells. We can see that such situations seem more or less plausible, but they are so profoundly divorced from human experience that we cannot comprehend them.

As with the event horizon of a black hole, there seems to be no possible communication between us and a Singularity. Our merely human reality is swallowed in an Einsteinian warp, and not so much as a single informative photon can creep back to us.

However, although the approach of a Vingean Singularity is easy to dramatize, it's not very relevant to what might really happen in a real posthuman world. Posthumans don't *care* if contemporary science fiction writers like myself or my esteemed colleague Dr. Vernor Vinge are able to properly imagine them. Our ignorance is not a constraint on them; it is merely a healthy measure of our human limitations.

Nevertheless, there are still a few ways for us to finesse the black hole of the Singularity, just as there are ways for us to map out good potential candidates for galactic black holes. Here are four things we can confidently say about this subject.

1. There is no one Singularity. Any area of scientific inquiry, pushed far enough, could provide its own native version of a transformative cataclysm: biological, cognitive, mechanical, cybernetic, and so on. If man is the measure of all things, then there is no measure by which we can't be made more than human. We might become ageless, or geniuses, or prosthetically enhanced, or cyberneticized, or any combination of those. We might be severely transformed by an unknown technology that we can't yet imagine.

2. A Singularity ends the human condition (because that is its definition), but it resolves nothing else. It would almost certainly be followed by a rapid, massive explosion of following Singularities. These ultracataclysmic events would disrupt the first Singularity even more than the first Singularity disrupted the original human condition. If we're living longer, then we aim for immortality; if we're superhumanly intelligent, all our possibilities explode in unforeseen directions. Posthumans aren't content with human achievements; they're better at posthumanity than we are.

3. The posthuman condition is banal. It is astounding, and eschatological, and ontological, but only by human standards. Oh, sure, we may become as gods (or something will, maybe a chimeric genetic hybrid or a genuinely smart computer), but the thrill fades fast, because that thrill is merely human and parochial. By the new, post-Singularity standards, posthumans are just as bored and frustrated as humans ever were. They are not magic, they are still quotidian entities in a gritty, rules-based physical universe. They will find themselves swiftly and bruisingly brought up against the limits of their own conditions, whatever those limits and conditions may be.

4. Messy, embarrassing, reversible, goofy, catch-as-catch-can posthumanism is politically preferable to sleek, streamlined, sudden, utter Final Solution posthumanism. The best way to encounter a Singularity would be to nick over the event horizon for a minute or two and have somebody else yank you back. Then the rest of us would be able to debrief you

and see just how far you would be able to jam that experience into language.

Lysergic acid, for instance, was loudly promoted as a life-changing, cosmic experience, but LSD, thankfully, lasts only eight hours in the system. This is well and good. If LSD were permanent, we'd be surrounded by many millions of middle-aged people who were irretrievably deranged.

And here is a fifth thing we might assert as well.

5. If it's hard to be a little bit pregnant, then it's even harder to be a little bit dead. Death, once known as the Great Leveler, will likely be rather less level than he used to be.

Once your metabolic function has ceased, you are not any more or less dead than our very dead friend Otzi. This dark understanding can be a source of great comfort. The dead are a great host, but each person among them died only one time. Dying one time is all the dying humans find necessary.

The traditional human response to death is to try to work through anger and rejection to acceptance. Typically, we humans imagine that we were put here by a higher power, to be taken away when it's our due time. A posthuman does not get the due time or the acceptance. He may have more command over his biological fate than we do, but that exacts a price. When his own "strange eventful history" is closed, that's not a kindness from the Grievous Angel; it's his own lookout.

It's a theoretical problem for us, a practical problem for them. Death is a grand abstraction, but it has to manifest itself as a physical process undergone by an individual. No one else can do your dying for you. Life extension, that cheery notion, comes with the sinister underside of "death extension." Death, that formerly brief, biologically necessary interregnum, can be extended to fantastic lengths far beyond any merely human experience. A posthuman runs the risk of spending an unconscionable amount of time in nonhuman states that are more or less alive.

Possibly, the authorities may appoint someone whose business it is to properly oversee your death. It's an open question whether you should ever trust any such person in such a vital, private, and personal matter. Professional euthanists don't seem likely to be any more capable or incorruptible than their current equivalents: doctors at the fatal bedside and judges trying capital crimes.

It's a situation that is poorly explored, a problem that the belle époque just doesn't get. It is tomorrow's problem, and we lack the habits and customs for it. Nor are we likely to create those habits and customs, at least not any solid, deep-rooted ones. A posthuman transformation will have a very hard time staying friendly with any human habit or custom. It will even have a lot of trouble with its own.

You may no longer be human, but you will not be divine or a superman. You will still have some kind of everyday treadmill, you will still have some quotidian engagement with time and space. Death will deserve respect.

This realization is the last and deepest futurist insight, the one that tows a true futurist out of the shallow waters of the huckster and the charlatan. The future is a lovely thing to contemplate, but in the final analysis, it is where we go to die. Shakespeare knew this, and since he was a great artist, he knew enough to mention the fact and then stop. His soliloquy had to end, just as this book does.

The world continues after our decease, and if we're lucky, we might go right on talking. But death is an event we can't finesse. Authentic futurism means staring directly into your own grave and the grave of everyone, and everything, you know and love.

Perhaps some words will help. The melancholy Jaques, my faithful guide within this book, was very big on words. Sermons in stones, and good in everything. Words are little enough to offer, but at the side of the gravestone, words are what we give. We might therefore imagine a useful document for the future, a kind of preobituary for the posthuman. It might look like this:

To Whom It May Concern:

When you read this, I will be dead. Forgive me for addressing you without allowing you any reply. Due to unique circumstances, I lacked an alternative.

I have passed from the scene of my own free will, by a method of my own choice that was thoroughly considered and appropriate to the desired end.

Unfortunately, due to the following objective realities [link to details of my pressing circumstances here], the continuation of my life has become untenable. I have therefore chosen this deliberate and final exit. This was no mere whim on my part, but a grave and solemn act whose circumstances were thoroughly considered. Please take note of my Last Will and Testament, a sound and sensible document that clearly demonstrates a balanced judgment on my part. This clear intent to smooth the future path of my heirs will speak for itself.

[Insert Last Will.]

Ending my life is my final tribute to its worthiness. Those who grieve for me should take comfort. The final moment is but one among many and cannot invalidate the others. Death is merely a necessity.

To prattle on after one's last will is surely an act in dubious taste. But it's a futurist's occupational hazard. In writing this book, I've rudely imposed myself into the emerging world of my children. That's a blatant embarrassment to all parties concerned, much like Dad crashing a teen dance with his rickety buck-and-wing. I've even sneaked furtively into the world of my grandchildren, who at this writing are entirely speculative entities. In those dodgy conditions, I'm a ghost who's writing to unborn phantoms. What's a ghost supposed to tell you people?

Once Grandpa has done his soft-shoe shuffle off his mortal coil, his party small talk slips from the merely undignified to the downright eldritch. Speaking from the grave can seem such a strange and piteous thing. Trust me; I know this from experience.

Whenever one visits a graveyard, attempting to commune with those worthy deceased, there's a powerful sense of calamity there. Victorian graveyards are pretty common in my corner of the planet. Those Victorians are not just *sort of* dead—why, they're *all* dead! The year 1880, for instance, has left us literally *no survivors,* not even one!

It's a simple matter to peruse those blurry stone exhortations of piety and sorrow. Rarely do you see an epitaph that really makes you want to know that grave's inhabitant any better. There's no human immediacy to these chiseled exhortations. They always fade into some blithering cult of ancestor worship—the temporal flip side of our peculiar regard for posterity. Quite commonly we just plain ignore our posterity. Sometimes we give it lip service. But when are we *friendly* with them? Nobody ever wants to *level with* posterity, to tell them something genuinely funny, gossipy, and deflating of pretense. It's all about the *judgment* of posterity— as if our descendants all wore black robes for a living and had nothing better to do with their own time than to laud and appreciate ours.

Victorian graveyards do infallibly offer me one stunning insight—that appalling number of dead children. Young wives dead in their teens and twenties, and a host of young men dead in plague and war. There's a scary rarity there of dead people over seventy. They dropped to earth like dragonflies in a cold snap.

But that is not at all the way Victorians saw themselves, even as they wept and dug those graves. On the contrary, one never saw an epoch more entirely progressive and self-congratulating. Very few peoples have ever been so pleased to be modern, so poised on top of the world. These people under the blurry tombstones—even the meekest and most downtrodden among them—knew they lived in a mighty age of fearless exploration, world conquest, and bold pioneering.

The most famous old grave in these parts belonged to a young woman we locals call the Leanderthal Lady. "Leanne" is a paleo-Texan ten thousand years old, several millennia the senior of Otzi. Her friends and kin buried her with a fossil shark tooth, which—I feel quite sure—must have been her favorite toy and totem. And compared with the awesome gulfs of time that fossil shark swam in—well, Leanne and I, and Otzi, and even that crowd of Victorians—we're all flirting and grinning at the same square dance, we're grooving under the mirror ball at the same discotheque.

A dead dragonfly is the creature of a mere season, frozen

in some dirty ice with those four mosaic wings and those curled and brittle legs. But in the private world that creature saw through its busy, faceted eyes, it was a flying tiger. It swam, killed, and ate, and then it flew, killed, and ate, and though its life is brief by the lumbering standards of us mammals, its genetic legacy is much older than ours, older even than dinosaurs'. Grief at its lot is our pathetic fallacy. We project our sorrow onto that potent little entity from within ourselves.

Some people are separated from us not by distance but by the ticking of the clock. A writer takes only debts and obligations from the past; he can tell them nothing. But if I may take the trouble here to speak to you, those who are to come, dear children of mine, as someone thoroughly dead—well, grief is just not a burden I would ever care to offer you. Given my druthers, I'd rather like us to dance, awkward though that may be.

Yes, yes, I'm your remote ancestor and a spiritual progenitor and all that; I can't and won't deny it, I'm sheepishly owning up. But I don't require you to trudge to my cenotaph in sackcloth and ashes. I'm not all demanding and judgmental about your bizarre and peculiar activities. On the contrary: I sympathize with your sense of inadequacy, your lasting awareness of hypocrisy and squalor. To the extent that such is possible—and it's always more possible than we think at first—I'd like us to have a down-and-dirty chumminess because of all that. We should aspire to a rich appreciation of one another's ironies and failings. Far from envying your stellar lot way up there in the glittering future, I'd love to help you celebrate your brief privilege of strutting the cosmic stage, of just plain being there-and-now. It's all right! You deserve it!

We dismiss the unborn as unknowable phantoms. We patronize the elder dead with the unearned benefits of hindsight. What we should strive to feel for both of them, for one another always and at all times, is a sense of sincere solidarity. If I win a role in your thoughts, unknown children, I would like to be a ghost you need not fear. A ghost whose strongest feelings for you are of unfeigned loving-kindness. Time may separate us, but outside yesterday and tomorrow, we can share the happy fact of being in the world.

AFTERWORD

TOMORROW THEN

Since the publication of this book about the future, the clock has continued to tick. The book left the presses and hit the shelves. I went on TV and radio, and I autographed a bunch of copies. Used ones started showing up. Gently, a fine layer of dust accumulated on my research documents. I got myself a nice new blobject computer.

So you and I are now dwelling in the future of this text. Now it's tomorrow. I'm rather happy with the way the future worked out for this book. The effort had its benefits. For instance, I got a nice, regular gig writing a futurist column for a big magazine. I have to watch the calendar for my deadline every month now, but you know, that clock-watching keeps me in shape.

You see, the very best way to become a futurist is to hang around until tomorrow shows up. Of course, once tomorrow becomes now, other people will probably dismiss that wonderful day as being merely "the present"—but, having read this book, you should know better than that. Today is the past's future. And man, whenever you look at today from the past's perspective, the future is something else.

The themes within this book don't yet look utterly outdated or weirdly far-fetched. They are pretty much bound to look that way, eventually. Some of these notions are becoming common wisdom. This encourages me. The most successful prognostications will become completely boring truisms, real clichés. This is a wondrous victory condition, rather like the remarkable fate of that Shakespeare guy. Just read one of those famous plays of his. Go ahead, try it! Why, it's just some stale bunch of quotes that everybody's already heard a hundred times!

The concept of New World Disorder is coming on strong. It used to sound quite DARPA-ish and RAND-y to speculate that the future of military conflict would be nation-states versus nonstate actors. But not anymore. Now it couldn't be more obvious that we're entering a newly bipolar world: the harried places with laws and plumbing versus the crazy places with gangs and guns.

The ominous part is the intimate way in which these two worlds globally intermingle. The Disorder has stolen a march. It took its war for chaos straight into New York City, NATO, and the UN, and it met with a signal success. Civilization is still reeling in rampant paranoia and astonishment. For the moment, that is. That won't last.

The political and infrastructural solutions to this challenge, should we manage to invent them, are going to be genuinely novel and profoundly interesting. Though I spend a whole lot of time looking for them, I truly have no idea what those global solutions will be. And neither does anyone else. The Great and the Good are not only bewildered by today's events, they're blatantly making up evidence to suit their preconceptions. This is a really interesting historical period. We are careening like a pinball, and we could go most anywhere.

In this brief Afterword, I want to offer some sound advice on how to persist long enough to see how things might be turning out in these dodgy circumstances. For a futurist, living long enough to become old-fashioned is the best revenge.

In circumstances like today's, we can borrow some useful tactical counsel from the U.S. Army Special Forces, who are a remarkably busy bunch of guys. These professionals are not in the business of remaining alive into the distant future, mind you. Instead, they're in the business of fulfilling their orders and obtaining the military objectives of their superiors while, if necessary, killing the enemy. But they do know some useful things about managing amid turbulence.

The great commonality that Special Forces and futurists

have is that they are both scouts by nature. It is their special business to go mess around in obscure, disturbing places of which normal people aren't yet taking much notice. With luck, they might come back and tell others.

The differences between Special Forces and futurists, however, abound. Futurists tend to be owlish, bookish, hyperrational types, while Special Forces are young, bold, brash guys in top physical condition. Futurists tend to emit works written in a hazy civilian blather, while Special Forces documents are militarily simple and direct. The guidelines for Special Forces are suitable for use by people in an almighty sweat, who might well be taking some incoming rounds. They're commonly phrased in blunt, sturdy acronyms and mnemonics.

When the Special Forces find themselves in unknown territory, disoriented, bewildered, and surrounded by possible or proven hostiles, they rely on a keyword, the acronym SURVIVAL, to help them remember their list of eight general survival principles:

1. (S)IZE UP THE SITUATION.
What exactly do you see around you? What condition are you yourself in? What kind of equipment do you have that might make a meaningful difference?

The goal here is to edge away from dazed bewilderment, and get some empirical referents. Don't just pray, hope, experience a sense of ominous foreboding, and sit down to bite your nails; go and measure something. Obtain some relevant facts. Make lists and charts, if you have to. Get a grip.

2. (U)NDUE HASTE MAKES WASTE.

Don't blow it in a panic in order to survive this moment. Assuming you do survive this moment, there's going to be another moment right on its heels. That moment will be just as urgent and awful, but you'll have wasted your resources. So be methodical. Even if you are losing your battle and stumbling in full retreat today, well, tomorrow is another day. And tomorrow has a tomorrow, too.

3. (R)EMEMBER WHERE YOU ARE.

You didn't get here by accident. This situation exists in some context of time and space. It has its causes and consequences. Keep the strategy in mind.

4. (V)ANQUISH FEAR AND PANIC.

You may be personally sweating it out on a battlefield—or maybe not. Nevertheless, to remain alive and alert will always require some output of willpower, discipline, and self-mastery. If you curl up, terrified, into a fetal ball, then a coffin makes a very natural next destination.

5. (I)MPROVISE.

Don't wait for orders from on high. Don't expect any perfect solutions. Try stuff out. Throw straws in the wind. Use models. Experiment. Place some small bets and observe the outcome. Tinker around.

6. (V)ALUE LIVING.

To stay alive successfully, you need to want that condition. This may sound simple-minded, but it is harder than it sounds. It can be hard to value living in the long-term, and fatally easy to valorize sudden death. In our dark era, the "martyrdom operation" has become the world's most potent political and military act.

People are mortal beings, born with an expiration date. Every human mind comes equipped with a death wish, which is part of our nature. Every soldier in every service has a streak of self-sacrifice. Every xenophobic has an element of destructive self-hatred.

People who die young tend to die from their own rash inexperience. But the people who die in middle age tend to die from dark fates that they brought upon themselves. They die from chronic self-insult. They may die rather gently and slowly from cigarettes, workaholism, overeating, or lack of exercise, but they do die. They die because they didn't value their lives enough to reform their beloved vices.

7. (A)CT LIKE THE NATIVES.

Do not despise and ignore the benighted people who have no idea what you are doing and thinking. You may not be there for the same reason that the natives are there. You certainly don't have to swallow their parochial ideology, their weird theology, or their senseless cultural values. Futurism is not a status contest between you and the mundane people. Futurism is an earnest attempt to achieve levels of cosmopolitan, transhistorical insight that are necessarily uncommon.

So do not think like the natives. Act like them. You should cultivate your ability to mimic the natives, to debrief them in a congenial way, and to adapt their practical habits when necessary. You have achieved something grand when you can act like the natives—and also cheerily act like whatever group it is that the natives ignore, fear, or shun most. At that point, you have truths to tell the natives that they have never allowed themselves to know.

8. (L)EARN BASIC SKILLS.

When you're facing a topic as big as All That Is to Come, it's good to have some professional methods. Personally, I favor journalism and industrial design, because they suit my temperament as a quick study who can coin some tag lines while hanging out next to cool hardware. But if you made a list of the futurists that I steal my best ideas from, you'd find physicists, economists, anthropologists, sociologists, business administrators, engineers, and, yes, even lawyers.

Such is the useful scheme for SURVIVAL within the Special Forces. Futurists, however, do face a few distinct occupational hazards that are not those of soldiers. Before I wrap up this Afterword, I must explore two of them. If you ever become a futurist, I think this counsel might help you.

1. DON'T WALK POINT.

Soldiers naturally have to do this, and the guy within the platoon who walks point gets shot at a whole lot. If you want to live a long time, you should not engage in this kind of activity without good reason.

It is not the core business of futurists to personally instigate massive social change. The people who gird their loins and rush out there to transform the world are not futurists. Basically, they are all mutants. These are people innately capable of extreme and perverse levels of performance. They are saints, prophets, gifted scientists, inventors, business moguls, revolutionaries, brilliant statesmen, and so forth and so on. They are commonly dangerous to themselves and to bystanders.

These world-changing people often have no idea that they are changing the world. If you tell them they are, it only irritates them. In their furiously focused daily lives, they are burrowing fanatically into some personal quirk or obsession, and, unlike futurists, they have very little interest in analyzing the bigger picture. People of this ilk exude social change in the way a plutonium rod exudes heat. World-changing pioneers should be treated with close attention and gingerly respect. Arrows fly around pioneers. Pioneers often get shot and scalped. Record-busting drag racers don't use their rearview mirrors much. They don't much care for brakes.

You, as a futurist, may understand what these people mean to the world and why they are important. This doesn't mean that you should try to become one of them. Be alert and watchful in their vicinity. Take notes. Analyze. Do not sell all your possessions and join their charismatic crusade. Do not suck up to them and expect them to make you rich. Do not fall in love with them and marry them and give them children. Do not, above all, get in their way and attempt to stop them.

If you want to really do these strange people a favor, show up in their days of despair and bleak ruin, when all their headlines have faded, and buy them a sandwich. You may well have a special power to cheer them up. Never try to do this by flattering them for their great historical achievement, though. If, for instance, your subject is a great world statesman who is in the deep political wilderness because he burned up all his political capital, you should not say to him, for instance, "Gosh, Mr. Gorbachev, thanks for the perestroika!" Instead, you should say something kind and supportive about his amateur weekend painting, or his choice in shoes, furniture, or folk music. Nobody ever appreciated these touchingly human aspects of his personality, because they've been overshadowed by the harsh klieg lights of his historical importance. World-changing people will be rather pleased when you tell them such a simple, homely thing;

you'll have done them a good turn. They will feel human. For them, that's a luxury.

Everybody gives to charity on Christmas. Futurists should beat the clock and give in February. Nobody much ever gives in February, but the cold wind and the hunger are every bit as keen then.

2. BEWARE THE MONKEY PUZZLE.

This legendary trap for wild monkeys involves planting a piece of candy bait inside a heavy coconut. The victim monkey snakes his clever fingers in and grips the tasty candy. Then he finds that he cannot withdraw his clenched fist from the narrow hole in the coconut's hard shell. Such is his romantic attraction to the candy prize, though, that he just cannot bear to release that sweet, yank his hand free, and flee for safety. Burdened by the weight and bulk of this coconut on his hairy, skinny arm, the monkey is easily run down and caught.

The lesson for futurists here is simple. Out in that wilderness of delightful, intriguing trends, conjectures, and happenstances is the one that you cannot resist. That is your Monkey Puzzle. It is the one futuristic curiosity that proves unbearably dear to your heart. It's the cause you shed hot tears over, the notion that you cherish for profound, irrational reasons. It is your wild hare, the scheme that you champion against all odds, the unlikelihood that you stoutly defend from a hostile and unbelieving world. Given its way, it will turn you from a scholar into a crank.

I have witnessed the Monkey Puzzle trapping many of my cherished colleagues. It is truly an occupational hazard. The Monkey Puzzle is rarely a major phenomenon. It is generally something rather trivial, silly, and goofy. You may find yourself longing to have your head frozen for millennia inside a tub of liquid nitrogen. You might be earnestly hunting Bigfoot with remote-controlled cameras. It might suddenly occur to you that, given all those decades of evidence and reports, UFOs might really and truly exist. The Monkey Puzzle is almost never based on a sober, rational analysis on the part of its victim. Instead, the Monkey Puzzle speaks to some underfed, sugar-starved part of the victim's personal psyche.

Sometimes the Monkey Puzzle is fatal. Generally, it just slows you down. But the Monkey Puzzle is very hard to guard against. To have one's hand stuck inside a coconut is disfiguring and renders you clumsy. To its victim, the Monkey Puzzle rarely looks or feels like a trap. The victim doesn't feel trapped; he feels heroic. The true Monkey Puzzle victim will hold his coconut-shrouded forearm in front of a mirror and say, "You know, on me, this thing looks great!"

Monkey Puzzles are generally visible to their victims only in retrospect: "What a fool I was." I have always wondered when I would get a Monkey Puzzle all my own. Only recently did I become pretty sure that I have already had, in my life to date, at least four of them. Luckily, they didn't last. And they came serially, instead of all at once. But you know, there's always tomorrow.

For a while there, I thought maybe writing this book was one of my Monkey Puzzles. Because it was so much fun and I enjoyed it so thoroughly, I worried that I might never stop typing.

But then, you know, eventually, one does.

ABOUT THE TYPE

This book was set in Minion, a 1990 Adobe Originals typeface by Robert Slimbach. Minion is inspired by classical, old-style typefaces of the late Renaissance, a period of elegant, beautiful, and highly readable type designs. Created primarily for text setting, Minion combines the aesthetic and functional qualities that make text type highly readable with versatility of digital technology.